A History of Science, V4

Henry Smith Williams

Table of Contents

A History of Science, V4

Henry Smith Williams

Kessinger Publishing reprints thousands of hard-to-find books!

Visit us at http://www.kessinger.net

A HISTORY OF SCIENCE
BY
HENRY SMITH WILLIAMS, M.D., LL.D.
ASSISTED BY
EDWARD H. WILLIAMS, M.D.

IN FIVE VOLUMES

BOOK IV. MODERN DEVELOPMENT OF THE CHEMICAL AND BIOLOGICAL SCIENCES

AS regards chronology, the epoch covered in the present volume is identical with that viewed in the preceding one. But now as regards subject matter we pass on to those diverse phases of the physical world which are the field of the chemist, and to those yet more intricate processes which have to do with living organisms. So radical are the changes here that we seem to be entering new worlds; and yet, here as before, there are intimations of the new discoveries away back in the Greek days. The solution of the problem of respiration will remind us that Anaxagoras half guessed the secret; and in those diversified studies which tell us of the Daltonian atom in its wonderful transmutations, we shall be reminded again of the Clazomenian philosopher and his successor Democritus.

Yet we should press the analogy much too far were we to intimate that the Greek of the elder day or any thinker of a more recent period had penetrated, even in the vaguest way, all of the mysteries that the nineteenth century has revealed in the fields of chemistry and biology. At the very most the insight of those great Greeks and of the wonderful seventeenth-century philosophers who so often seemed on the verge of our later discoveries did no more than vaguely anticipate their successors of this later century. To gain an accurate, really specific knowledge of the properties of elementary bodies was reserved for the chemists of a recent epoch. The vague Greek questionings as to organic evolution were world-wide from the precise inductions of a Darwin. If the mediaeval Arabian endeavored to dull the knife of the surgeon with the use of drugs, his results hardly merit to be termed even an anticipation of modern anaesthesia. And when we speak of preventive medicine—of bacteriology in all its phases—we have to do with a marvellous field of which no previous generation of men had even the slightest inkling.

All in all, then, those that lie before us are perhaps the most wonderful and the most fascinating of all the fields of science. As the chapters of the preceding book carried us out into a macrocosm of inconceivable magnitude, our present studies are to reveal a microcosm of equally inconceivable smallness. As the studies of the physicist attempted to reveal the very nature of matter and of energy, we have now to seek the solution of the yet more inscrutable problems of life and of mind.

I. THE PHLOGISTON THEORY IN CHEMISTRY

The development of the science of chemistry from the "science" of alchemy is a striking example of the complete revolution in the attitude of observers in the field of science. As has been pointed out in a preceding chapter, the alchemist, having a preconceived idea of how things should be, made all his experiments to prove his preconceived theory; while the chemist reverses this attitude of mind and bases his conceptions on the results of his laboratory experiments. In short, chemistry is what alchemy never could be, an inductive science. But this transition from one point of view to an exactly opposite one was necessarily a very slow process. Ideas that have held undisputed sway over the minds of succeeding generations for hundreds of years cannot be overthrown in a moment, unless the agent of such an overthrow be so obvious that it cannot be challenged. The rudimentary chemistry that overthrew alchemy had nothing so obvious and palpable.

The great first step was the substitution of the one principle, phlogiston, for the three principles, salt, sulphur, and mercury. We have seen how the experiment of burning or calcining such a metal as lead "destroyed" the lead as such, leaving an entirely different substance in its place, and how the original metal could be restored by the addition of wheat to the calcined product. To the alchemist this was "mortification" and "revivification" of the metal. For, as pointed out by Paracelsus, "anything that could be killed by man could also be revivified by him, although this was not possible to the things killed by God." The burning of such substances as wood, wax, oil, etc., was also looked upon as the same "killing" process, and the fact that the alchemist was unable to revivify them was regarded as simply the lack of skill on his part, and in no wise affecting the theory itself.

But the iconoclastic spirit, if not the acceptance of all the teachings, of the great Paracelsus had been gradually taking root among the better class of alchemists, and about the middle of the seventeenth century Robert Boyle (1626–1691) called attention to the possibility of making a wrong deduction from the phenomenon of the calcination of the metals, because of a very important factor, the action of the air, which was generally overlooked. And he urged his colleagues of the laboratories to give greater heed to certain other phenomena that might pass unnoticed in the ordinary calcinating process. In his work, The Sceptical Chemist, he showed the reasons for doubting the threefold constitution of matter; and in his General History of the Air advanced some novel and carefully studied theories as to the composition of the atmosphere. This was an important step, and although Boyle is not directly responsible for the phlogiston theory, it is probable that his experiments on the atmosphere influenced considerably the real founders, Becker and Stahl.

Boyle gave very definitely his idea of how he thought air might be composed. "I conjecture that the atmospherical air consists of three different kinds of corpuscles," he says; "the first, those numberless particles which, in the form of vapors or dry exhalations, ascend from the earth, water, minerals, vegetables, animals, etc.; in a word, whatever substances are elevated by the celestial or subterraneal heat, and thence diffused into the atmosphere. The second may be yet more subtle, and consist of those exceedingly minute atoms, the magnetical effluvia of the earth, with other innumerable particles sent out from the bodies of the celestial luminaries, and causing, by their influence, the idea of light in us. The third sort is its characteristic and essential property, I mean permanently elastic parts. Various hypotheses may be framed relating to the structure of these later particles of the air. They might be resembled to the springs of watches, coiled up and endeavoring to restore themselves; to wool, which, being compressed, has an elastic force; to slender wires of different substances, consistencies, lengths, and thickness; in greater curls or less, near to, or remote from each other, etc., yet all continuing springy, expansible, and compressible. Lastly, they may also be compared to the thin shavings of different kinds of wood, various in their lengths, breadth, and thickness. And this, perhaps, will seem the most eligible hypothesis, because it, in some measure, illustrates the production of the elastic particles we are considering. For no art or curious instruments are required to make these shavings whose curls are in no wise uniform, but seemingly casual; and what is more remarkable, bodies that before seemed unelastic, as beams and blocks, will afford them."[1]

Although this explanation of the composition of the air is most crude, it had the effect of directing attention to the fact that the atmosphere is not "mere nothingness," but a "something" with a definite composition, and this served as a good foundation for future investigations. To be sure, Boyle was neither the first nor the only chemist who had suspected that the air was a mixture of gases, and not a simple one, and that only certain of these gases take part in the process of calcination. Jean Rey, a French physician, and John Mayow, an Englishman, had preformed experiments which showed conclusively that the air was not a simple substance; but Boyle's work was better known, and in its effect probably more important. But with all Boyle's explanations of the composition of air, he still believed that there was an inexplicable something, a "vital substance," which he was unable to fathom, and which later became the basis of Stahl's phlogiston theory. Commenting on this mysterious substance, Boyle says: "The, difficulty we find in keeping flame and fire alive, though but for a little time, without air, renders it suspicious that there be dispersed through the rest of the atmosphere some odd substance, either of a solar, astral, or other foreign nature; on account of which the air is so necessary to the substance of flame!" It was this idea that attracted the attention of George Ernst Stahl (1660–1734), a professor of medicine in the University of Halle, who later founded his new theory upon it. Stahl's theory was a development of an earlier chemist, Johann Joachim Becker (1635–1682), in whose footsteps he followed and whose experiments he carried further.

In many experiments Stahl had been struck with the fact that certain substances, while differing widely, from one another in many respects, were alike in combustibility. From this he argued that all combustible substances must contain a common principle, and this principle he named phlogiston. This phlogiston he believed to be intimately associated in combination with other substances in nature, and in that condition not perceivable by the senses; but it was supposed to escape as a substance burned, and become apparent to the senses as fire or flame. In other words, phlogiston was something imprisoned in a combustible structure (itself forming part of the structure), and only liberated when this structure was destroyed. Fire, or flame, was FREE phlogiston, while the imprisoned phlogiston was called COMBINED PHLOGISTON, or combined fire. The peculiar quality of this strange substance was that it disliked freedom and was always striving to conceal itself in some combustible substance. Boyle's tentative suggestion that heat was simply motion was apparently not accepted by Stahl, or perhaps it was unknown to him.

According to the phlogistic theory, the part remaining after a substance was burned was simply the original substance deprived of phlogiston. To restore the original combustible substance, it was necessary to heat the residue of the combustion with something that burned easily, so that the freed phlogiston might again combine with the ashes. This was explained by the supposition that the more combustible a substance was the more phlogiston it contained, and since free phlogiston sought always to combine with some suitable substance, it was only necessary to mix the phlogisticating agents, such as charcoal, phosphorus, oils, fats, etc., with the ashes of the original substance, and heat the mixture, the phlogiston thus freed uniting at once with the ashes. This theory fitted very nicely as applied to the calcined lead revivified by the grains of wheat, although with some other products of calcination it did not seem to apply at all.

It will be seen from this that the phlogistic theory was a step towards chemistry and away from alchemy. It led away from the idea of a "spirit" in metals that could not be seen, felt, or appreciated by any of the senses, and substituted for it a principle which, although a falsely conceived one, was still much more tangible than the "spirit," since it could be seen and felt as free phlogiston and weighed and measured as combined phlogiston. The definiteness of the statement that a metal, for example, was composed of phlogiston and an element was much less enigmatic, even if wrong, than the statement of the alchemist that "metals are produced by the spiritual action of the three principles, salt, mercury, sulphur"—particularly when it is explained that salt, mercury, and sulphur were really not what their names implied, and that there was no universally accepted belief as to what they really were.

The metals, which are now regarded as elementary bodies, were considered compounds by the phlogistians, and they believed that the calcining of a metal was a process of simplification. They noted, however, that the remains of calcination weighed more than the original product, and the natural inference from this would be that the metal must have taken in some substance rather than have given off anything. But the phlogistians had not learned the all−important significance of weights, and their explanation of variation in weight was either that such gain or loss was an unimportant "accident" at best, or that phlogiston, being light, tended to lighten any substance containing it, so that driving it out of the metal by calcination naturally left the residue heavier.

At first the phlogiston theory seemed to explain in an indisputable way all the known chemical phenomena. Gradually, however, as experiments multiplied, it became evident that the plain theory as stated by Stahl and his followers failed to explain satisfactorily certain laboratory reactions. To meet these new conditions, certain modifications were introduced from time to time, giving the theory a flexibility that would allow it to cover all cases. But as the number of inexplicable experiments continued to increase, and new modifications to the theory became necessary, it was found that some of these modifications were directly contradictory to others, and thus the simple theory became too cumbersome from the number of its modifications. Its supporters disagreed among themselves, first as to the explanation of certain phenomena that did not seem to accord with the phlogistic theory, and a little later as to the theory itself. But as yet there was no satisfactory substitute for this theory, which, even if unsatisfactory, seemed better than anything that had gone before or could be suggested.

But the good effects of the era of experimental research, to which the theory of Stahl had given such an impetus, were showing in the attitude of the experimenters. The works of some of the older writers, such as Boyle and Hooke, were again sought out in their dusty corners and consulted, and their surmises as to the possible mixture of various gases in the air were more carefully considered. Still the phlogiston theory was firmly grounded in the minds of the philosophers, who can hardly be censured for adhering to it, at least until some satisfactory substitute was offered. The foundation for such a theory was finally laid, as we shall see presently, by the work of Black, Priestley, Cavendish, and Lavoisier, in the eighteenth century, but the phlogiston theory cannot be said to have finally succumbed until the opening years of the nineteenth century.

II. THE BEGINNINGS OF MODERN CHEMISTRY

THE "PNEUMATIC" CHEMISTS

Modern chemistry may be said to have its beginning with the work of Stephen Hales (1677–1761), who early in the eighteenth century began his important study of the elasticity of air. Departing from the point of view of most of the scientists of the time, be considered air to be "a fine elastic fluid, with particles of very different nature floating in it" ; and he showed that these "particles" could be separated. He pointed out, also, that various gases, or "airs," as he called them, were contained in many solid substances. The importance of his work, however, lies in the fact that his general studies were along lines leading away from the accepted doctrines of the time, and that they gave the impetus to the investigation of the properties of gases by such chemists as Black, Priestley, Cavendish, and Lavoisier, whose specific discoveries are the foundation–stones of modern chemistry.

JOSEPH BLACK

The careful studies of Hales were continued by his younger confrere, Dr. Joseph Black (1728–1799), whose experiments in the weights of gases and other chemicals were first steps in quantitative chemistry. But even more important than his discoveries of chemical properties in general was his discovery of the properties of carbonic–acid gas.

Black had been educated for the medical profession in the University of Glasgow, being a friend and pupil of the famous Dr. William Cullen. But his liking was for the chemical laboratory rather than for the practice of medicine. Within three years after completing his medical course, and when only twenty-three years of age, he made the discovery of the properties of carbonic acid, which he called by the name of "fixed air." After discovering this gas, Black made a long series of experiments, by which he was able to show how widely it was distributed throughout nature. Thus, in 1757, be discovered that the bubbles given off in the process of brewing, where there was vegetable fermentation, were composed of it. To prove this, he collected the contents of these bubbles in a bottle containing lime-water. When this bottle was shaken violently, so that the lime-water and the carbonic acid became thoroughly mixed, an insoluble white powder was precipitated from the solution, the carbonic acid having combined chemically with the lime to form the insoluble calcium carbonate, or chalk. This experiment suggested another. Fixing a piece of burning charcoal in the end of a bellows, he arranged a tube so that the gas coming from the charcoal would pass through the lime-water, and, as in the case of the bubbles from the brewer's vat, he found that the white precipitate was thrown down; in short, that carbonic acid was given off in combustion. Shortly after, Black discovered that by blowing through a glass tube inserted into lime-water, chalk was precipitated, thus proving that carbonic acid was being constantly thrown off in respiration.

The effect of Black's discoveries was revolutionary, and the attitude of mind of the chemists towards gases, or "airs," was changed from that time forward. Most of the chemists, however, attempted to harmonize the new facts with the older theories—to explain all the phenomena on the basis of the phlogiston theory, which was still dominant. But while many of Black's discoveries could not be made to harmonize with that theory, they did not directly overthrow it. It required the additional discoveries of some of Black's fellow-scientists to complete its downfall, as we shall see.

HENRY CAVENDISH

This work of Black's was followed by the equally important work of his former pupil, Henry Cavendish (1731-1810), whose discovery of the composition of many substances, notably of nitric acid and of water, was of great importance, adding another link to the important chain of evidence against the phlogiston theory. Cavendish is one of the most eccentric figures in the history of science, being widely known in his own time for his immense wealth and brilliant intellect, and also for his peculiarities and his morbid sensibility, which made him dread society, and probably did much in determining his career. Fortunately for him, and incidentally for the cause of science, he was able to pursue laboratory investigations without being obliged to mingle with his dreaded fellow—mortals, his every want being provided for by the immense fortune inherited from his father and an uncle.

When a young man, as a pupil of Dr. Black, he had become imbued with the enthusiasm of his teacher, continuing Black's investigations as to the properties of carbonic—acid gas when free and in combination. One of his first investigations was reported in 1766, when he communicated to the Royal Society his experiments for ascertaining the properties of carbonic—acid and hydrogen gas, in which he first showed the possibility of weighing permanently elastic fluids, although Torricelli had before this shown the relative weights of a column of air and a column of mercury. Other important experiments were continued by Cavendish, and in 1784 he announced his discovery of the composition of water, thus robbing it of its time—honored position as an "element." But his claim to priority in this discovery was at once disputed by his fellow—countryman James Watt and by the Frenchman Lavoisier. Lavoisier's claim was soon disallowed even by his own countrymen, but for many years a bitter controversy was carried on by the partisans of Watt and Cavendish. The two principals, however, seem. never to have entered into this controversy with anything like the same ardor as some of their successors, as they remained on the best of terms.[1] It is certain, at any rate, that Cavendish announced his discovery officially before Watt claimed that the announcement had been previously made by him, "and, whether right or wrong, the honor of scientific discoveries seems to be accorded naturally to the man who first publishes a demonstration of his discovery." Englishmen very generally admit the justness of Cavendish's claim, although the French scientist Arago, after reviewing the evidence carefully in 1833, decided in favor of Watt.

It appears that something like a year before Cavendish made known his complete demonstration of the composition of water, Watt communicated to the Royal Society a suggestion that water was composed of "dephlogisticated air (oxygen) and phlogiston (hydrogen) deprived of part of its latent heat." Cavendish knew of the suggestion, but in his experiments refuted the idea that the hydrogen lost any of its latent heat. Furthermore, Watt merely suggested the possible composition without proving it, although his idea was practically correct, if we can rightly interpret the vagaries of the nomenclature then in use. But had Watt taken the steps to demonstrate his theory, the great "Water Controversy" would have been avoided. Cavendish's report of his discovery to the Royal

Society covers something like forty pages of printed matter. In this he shows how, by passing an electric spark through a closed jar containing a mixture of hydrogen gas and oxygen, water is invariably formed, apparently by the union of the two gases. The experiment was first tried with hydrogen and common air, the oxygen of the air uniting with the hydrogen to form water, leaving the nitrogen of the air still to be accounted for. With pure oxygen and hydrogen, however, Cavendish found that pure water was formed, leaving slight traces of any other, substance which might not be interpreted as being Chemical impurities. There was only one possible explanation of this phenomenon—that hydrogen and oxygen, when combined, form water.

"By experiments with the globe it appeared," wrote Cavendish, "that when inflammable and common air are exploded in a proper proportion, almost all the inflammable air, and near one–fifth the common air, lose their elasticity and are condensed into dew. And by this experiment it appears that this dew is plain water, and consequently that almost all the inflammable air is turned into pure water.

"In order to examine the nature of the matter condensed on firing a mixture of dephlogisticated and inflammable air, I took a glass globe, holding 8800 grain measures, furnished with a brass cock and an apparatus for firing by electricity. This globe was well exhausted by an air–pump, and then filled with a mixture of inflammable and dephlogisticated air by shutting the cock, fastening the bent glass tube into its mouth, and letting up the end of it into a glass jar inverted into water and containing a mixture of 19,500 grain measures of dephlogisticated air, and 37,000 of inflammable air; so that, upon opening the cock, some of this mixed air rushed through the bent tube and filled the globe. The cock was then shut and the included air fired by electricity, by means of which almost all of it lost its elasticity (was condensed into water vapors). The cock was then again opened so as to let in more of the same air to supply the place of that destroyed by the explosion, which was again fired, and the operation continued till almost the whole of the mixture was let into the globe and exploded. By this means, though the globe held not more than a sixth part of the mixture, almost the whole of it was exploded therein without any fresh exhaustion of the globe."

At first this condensed matter was "acid to the taste and contained two grains of nitre," but Cavendish, suspecting that this was due to impurities, tried another experiment that proved conclusively that his opinions were correct. "I therefore made another experiment," he says, "with some more of the same air from plants in which the proportion of inflammable air was greater, so that the burnt air was almost completely phlogisticated, its standard being one–tenth. The condensed liquor was then not at all acid, but seemed pure water."

From these experiments he concludes "that when a mixture of inflammable and dephlogisticated air is exploded, in such proportions that the burnt air is not much phlogisticated, the condensed liquor contains a little acid which is always of the nitrous kind, whatever substance the dephlogisticated air is procured from; but if the proportion be such that the burnt air is almost entirely phlogisticated, the condensed liquor is not at all acid, but seems pure water, without any addition whatever."[2]

These same experiments, which were undertaken to discover the composition of water, led him to discover also the composition of nitric acid. He had observed that, in the combustion of hydrogen gas with common air, the water was slightly tinged with acid, but that this was not the case when pure oxygen gas was used. Acting upon this observation, he devised an experiment to determine the nature of this acid. He constructed an apparatus whereby an electric spark was passed through a vessel containing common air. After this process had been carried on for several weeks a small amount of liquid was formed. This liquid combined with a solution of potash to form common nitre, which "detonated with charcoal, sparkled when paper impregnated with it was burned, and gave out nitrous fumes when sulphuric acid was poured on it." In other words, the liquid was shown to be nitric acid. Now, since nothing but pure air had been used in the initial experiment, and since air is composed of nitrogen and oxygen, there seemed no room to doubt that nitric acid is a combination of nitrogen and oxygen.

This discovery of the nature of nitric acid seems to have been about the last work of importance that Cavendish did in the field of chemistry, although almost to the hour of his death he was constantly occupied with scientific observations. Even in the last moments of his life this habit asserted itself, according to Lord Brougham. "He died on March 10, 1810, after a short illness, probably the first, as well as the last, which he ever suffered. His habit of curious observation continued to the end. He was desirous of marking the progress of the disease and the gradual extinction of the vital powers. With these ends in view, that he might not be disturbed, he desired to be left alone. His servant, returning sooner than he had wished, was ordered again to leave the chamber of death, and when be came back a second time he found his master had expired.[3]

JOSEPH PRIESTLEY

While the opulent but diffident Cavendish was making his important discoveries, another Englishman, a poor country preacher named Joseph Priestley (1733–1804) was not only rivalling him, but, if anything, outstripping him in the pursuit of chemical discoveries. In 1761 this young minister was given a position as tutor in a nonconformist academy at Warrington, and here, for six years, he was able to pursue his studies in chemistry and electricity. In 1766, while on a visit to London, he met Benjamin Franklin, at whose suggestion he published his History of Electricity. From this time on he made steady progress in scientific investigations, keeping up his ecclesiastical duties at the same time. In 1780 he removed to Birmingham, having there for associates such scientists as James Watt, Boulton, and Erasmus Darwin.

Eleven years later, on the anniversary of the fall of the Bastile in Paris, a fanatical mob, knowing Priestley's sympathies with the French revolutionists, attacked his house and chapel, burning both and destroying a great number of valuable papers and scientific instruments. Priestley and his family escaped violence by flight, but his most cherished possessions were destroyed; and three years later he quitted England forever, removing to the United States, whose struggle for liberty he had championed. The last ten years of his life were spent at Northumberland, Pennsylvania, where he continued his scientific researches.

Early in his scientific career Priestley began investigations upon the "fixed air" of Dr. Black, and, oddly enough, he was stimulated to this by the same thing that had influenced Black—that is, his residence in the immediate neighborhood of a brewery. It was during the course of a series of experiments on this and other gases that he made his greatest discovery, that of oxygen, or "dephlogisticated air," as he called it. The story of this important discovery is probably best told in Priestley's own words:

"There are, I believe, very few maxims in philosophy that have laid firmer hold upon the mind than that air, meaning atmospheric air, is a simple elementary substance, indestructible and unalterable, at least as much so as water is supposed to be. In the course of my inquiries I was, however, soon satisfied that atmospheric air is not an unalterable thing; for that, according to my first hypothesis, the phlogiston with which it becomes loaded from bodies burning in it, and the animals breathing it, and various other chemical processes, so far alters and depraves it as to render it altogether unfit for inflammation, respiration, and other purposes to which it is subservient; and I had discovered that agitation in the water, the process of vegetation, and probably other natural processes, restore it to its original purity....

"Having procured a lens of twelve inches diameter and twenty inches local distance, I proceeded with the greatest alacrity, by the help of it, to discover what kind of air a great variety of substances would yield, putting them into the vessel, which I filled with quicksilver, and kept inverted in a basin of the same .... With this apparatus, after a variety of experiments .... on the 1st of August, 1774, I endeavored to extract air from mercurius calcinatus per se; and I presently found that, by means of this lens, air was expelled from it very readily. Having got about three or four times as much as the bulk of my materials, I admitted water to it, and found that it was not imbibed by it. But what surprised me more than I can express was that a candle burned in this air with a remarkably vigorous flame, very much like that enlarged flame with which a candle burns in nitrous oxide, exposed to iron or liver of sulphur; but as I had got nothing like this remarkable appearance from any kind of air besides this particular modification of vitrous air, and I knew no vitrous acid was used in the preparation of mercurius calcinatus, I was utterly at a loss to account for it."[4]

The "new air" was, of course, oxygen. Priestley at once proceeded to examine it by a long series of careful experiments, in which, as will be seen, he discovered most of the remarkable qualities of this gas. Continuing his description of these experiments, he says:

"The flame of the candle, besides being larger, burned with more splendor and heat than in that species of nitrous air; and a piece of red−hot wood sparkled in it, exactly like paper dipped in a solution of nitre, and it consumed very fast; an experiment that I had never thought of trying with dephlogisticated nitrous air.

". . . I had so little suspicion of the air from the mercurius calcinatus, etc., being wholesome, that I had not even thought of applying it to the test of nitrous air; but thinking (as my reader must imagine I frequently must have done) on the candle burning in it after long agitation in water, it occurred to me at last to make the experiment; and, putting one measure of nitrous air to two measures of this air, I found not only that it was diminished, but that it was diminished quite as much as common air, and that the redness of the mixture was likewise equal to a similar mixture of nitrous and common air.... The next day I was more surprised than ever I had been before with finding that, after the above-mentioned mixture of nitrous air and the air from mercurius calcinatus had stood all night, . . . a candle burned in it, even better than in common air."

A little later Priestley discovered that "dephlogisticated air . . . is a principal element in the composition of acids, and may be extracted by means of heat from many substances which contain them.... It is likewise produced by the action of light upon green vegetables; and this seems to be the chief means employed to preserve the purity of the atmosphere."

This recognition of the important part played by oxygen in the atmosphere led Priestley to make some experiments upon mice and insects, and finally upon himself, by inhalations of the pure gas. "The feeling in my lungs," he said, "was not sensibly different from that of common air, but I fancied that my breathing felt peculiarly light and easy for some time afterwards. Who can tell but that in time this pure air may become a fashionable article in luxury? . . . Perhaps we may from these experiments see that though pure dephlogisticated air might be useful as a medicine, it might not be so proper for us in the usual healthy state of the body."

This suggestion as to the possible usefulness of oxygen as a medicine was prophetic. A century later the use of oxygen had become a matter of routine practice with many physicians. Even in Priestley's own time such men as Dr. John Hunter expressed their belief in its efficacy in certain conditions, as we shall see, but its value in medicine was not fully appreciated until several generations later.

Several years after discovering oxygen Priestley thus summarized its properties: "It is this ingredient in the atmospheric air that enables it to support combustion and animal life. By means of it most intense heat may be produced, and in the purest of it animals will live nearly five times as long as in an equal quantity of atmospheric air. In respiration, part of this air, passing the membranes of the lungs, unites with the blood and imparts to it its florid color, while the remainder, uniting with phlogiston exhaled from venous blood, forms mixed air. It is dephlogisticated air combined with water that enables fishes to live in it."[5]

KARL WILHELM SCHEELE

The discovery of oxygen was the last but most important blow to the tottering phlogiston theory, though Priestley himself would not admit it. But before considering the final steps in the overthrow of Stahl's famous theory and the establishment of modern chemistry, we must review the work of another great chemist, Karl Wilhelm Scheele (1742–1786), of Sweden, who discovered oxygen quite independently, although later than Priestley. In the matter of brilliant discoveries in a brief space of time Scheele probably eclipsed all his great contemporaries. He had a veritable genius for interpreting chemical reactions and discovering new substances, in this respect rivalling Priestley himself. Unlike Priestley, however, he planned all his experiments along the lines of definite theories from the beginning, the results obtained being the logical outcome of a predetermined plan.

Scheele was the son of a merchant of Stralsund, Pomerania, which then belonged to Sweden. As a boy in school he showed so little aptitude for the study of languages that he was apprenticed to an apothecary at the age of fourteen. In this work he became at once greatly interested, and, when not attending to his duties in the dispensary, he was busy day and night making experiments or studying books on chemistry. In 1775, still employed as an apothecary, he moved to Stockholm, and soon after he sent to Bergman, the leading chemist of Sweden, his first discovery—that of tartaric acid, which he had isolated from cream of tartar. This was the beginning of his career of discovery, and from that time on until his death he sent forth accounts of new discoveries almost uninterruptedly. Meanwhile he was performing the duties of an ordinary apothecary, and struggling against poverty. His treatise upon Air and Fire appeared in 1777. In this remarkable book he tells of his discovery of oxygen—"empyreal" or "fire–air," as he calls it—which he seems to have made independently and without ever having heard of the previous discovery by Priestley. In this book, also, he shows that air is composed chiefly of oxygen and nitrogen gas.

Early in his experimental career Scheele undertook the solution of the composition of black oxide of manganese, a substance that had long puzzled the chemists. He not only succeeded in this, but incidentally in the course of this series of experiments he discovered oxygen, baryta, and chlorine, the last of far greater importance, at least commercially, than the real object of his search. In speaking of the experiment in which the discovery was made he says:

"When marine (hydrochloric) acid stood over manganese in the cold it acquired a dark reddish–brown color. As manganese does not give any colorless solution without uniting with phlogiston [probably meaning hydrogen], it follows that marine acid can dissolve it without this principle. But such a solution has a blue or red color. The color is here more brown than red, the reason being that the very finest portions of the manganese, which do not sink so easily, swim in the red solution; for without these fine particles the solution is red, and red mixed with black is brown. The manganese has here attached itself so loosely to acidum salis that the water can precipitate it, and this precipitate behaves like ordinary manganese. When, now, the mixture of manganese and spiritus salis was set to digest, there arose an effervescence and smell of aqua regis."[6]

The "effervescence" he refers to was chlorine, which he proceeded to confine in a suitable vessel and examine more fully. He described it as having a "quite characteristically suffocating smell," which was very offensive. He very soon noted the decolorizing or bleaching effects of this now product, finding that it decolorized flowers, vegetables, and many other substances.

Commercially this discovery of chlorine was of enormous importance, and the practical application of this new chemical in bleaching cloth soon supplanted the, old process of crofting—that is, bleaching by spreading the cloth upon the grass. But although Scheele first pointed out the bleaching quality of his newly discovered gas, it was the French savant, Berthollet, who, acting upon Scheele's discovery that the new gas would decolorize vegetables and flowers, was led to suspect that this property might be turned to account in destroying the color of cloth. In 1785 he read a paper before the Academy of Sciences of Paris, in which he showed that bleaching by chlorine was entirely satisfactory, the color but not the substance of the cloth being affected. He had experimented previously and found that the chlorine gas was soluble in water and could thus be made practically available for bleaching purposes. In 1786 James Watt examined specimens of the bleached cloth made by Berthollet, and upon his return to England first instituted the process of practical bleaching. His process, however, was not entirely satisfactory, and, after undergoing various modifications and improvements, it was finally made thoroughly practicable by Mr. Tennant, who hit upon a compound of chlorine and lime—the chloride of lime—which was a comparatively cheap chemical product, and answered the purpose better even than chlorine itself.

To appreciate how momentous this discovery was to cloth manufacturers, it should be remembered that the old process of bleaching consumed an entire summer for the whitening of a single piece of linen; the new process reduced the period to a few hours.

To be sure, lime had been used with fair success previous to Tennant's discovery, but successful and practical bleaching by a solution of chloride of lime was first made possible by him and through Scheele's discovery of chlorine.

Until the time of Scheele the great subject of organic chemistry had remained practically unexplored, but under the touch of his marvellous inventive genius new methods of isolating and studying animal and vegetable products were introduced, and a large number of acids and other organic compounds prepared that had been hitherto unknown. His explanations of chemical phenomena were based on the phlogiston theory, in which, like Priestley, he always, believed. Although in error in this respect, he was, nevertheless, able to make his discoveries with extremely accurate interpretations. A brief epitome of the list of some of his more important discoveries conveys some idea, of his fertility of mind as well as his industry. In 1780 he discovered lactic acid,[7] and showed that it was the substance that caused the acidity of sour milk; and in the same year he discovered mucic acid. Next followed the discovery of tungstic acid, and in 1783 he added to his list of useful discoveries that of glycerine. Then in rapid succession came his announcements of the new vegetable products citric, malic, oxalic, and gallic acids. Scheele not only made the discoveries, but told the world how he had made them—how any chemist might have made them if he chose—for he never considered that he had really discovered any substance until he had made it, decomposed it, and made it again.

His experiments on Prussian blue are most interesting, not only because of the enormous amount of work involved and the skill he displayed in his experiments, but because all the time the chemist was handling, smelling, and even tasting a compound of one of the most deadly poisons, ignorant of the fact that the substance was a dangerous one to handle. His escape from injury seems almost miraculous; for his experiments, which were most elaborate, extended over a considerable period of time, during which he seems to have handled this chemical with impunity.

While only forty years of age and just at the zenith of his fame, Scheele was stricken by a fatal illness, probably induced by his ceaseless labor and exposure. It is gratifying to know, however, that during the last eight or nine years of his life he had been less bound down by pecuniary difficulties than before, as Bergman had obtained for him an annual grant from the Academy. But it was characteristic of the man that, while devoting one−sixth of the amount of this grant to his personal wants, the remaining five−sixths was devoted to the expense of his experiments.

LAVOISIER AND THE FOUNDATION OF MODERN CHEMISTRY

The time was ripe for formulating the correct theory of chemical composition: it needed but the master hand to mould the materials into the proper shape. The discoveries in chemistry during the eighteenth century had been far-reaching and revolutionary in character. A brief review of these discoveries shows how completely they had subverted the old ideas of chemical elements and chemical compounds. Of the four substances earth, air, fire, and water, for many centuries believed to be elementary bodies, not one has stood the test of the eighteenth-century chemists. Earth had long since ceased to be regarded as an element, and water and air had suffered the same fate in this century. And now at last fire itself, the last of the four "elements" and the keystone to the phlogiston arch, was shown to be nothing more than one of the manifestations of the new element, oxygen, and not "phlogiston" or any other intangible substance.

In this epoch of chemical discoveries England had produced such mental giants and pioneers in science as Black, Priestley, and Cavendish; Sweden had given the world Scheele and Bergman, whose work, added to that of their English confreres, had laid the broad base of chemistry as a science; but it was for France to produce a man who gave the final touches to the broad but rough workmanship of its foundation, and establish it as the science of modern chemistry. It was for Antoine Laurent Lavoisier (1743-1794) to gather together, interpret correctly, rename, and classify the wealth of facts that his immediate predecessors and contemporaries had given to the world.

The attitude of the mother-countries towards these illustrious sons is an interesting piece of history. Sweden honored and rewarded Scheele and Bergman for their efforts; England received the intellectuality of Cavendish with less appreciation than the Continent, and a fanatical mob drove Priestley out of the country; while France, by sending Lavoisier to the guillotine, demonstrated how dangerous it was, at that time at least, for an intelligent Frenchman to serve his fellowman and his country well.

"The revolution brought about by Lavoisier in science," says Hoefer, "coincides by a singular act of destiny with another revolution, much greater indeed, going on then in the political and social world. Both happened on the same soil, at the same epoch, among the same people; and both marked the commencement of a new era in their respective spheres."[8]

Lavoisier was born in Paris, and being the son of an opulent family, was educated under the instruction of the best teachers of the day. With Lacaille he studied mathematics and astronomy; with Jussieu, botany; and, finally, chemistry under Rouelle. His first work of importance was a paper on the practical illumination of the streets of Paris, for which a prize had been offered by M. de Sartine, the chief of police. This prize was not awarded to Lavoisier, but his suggestions were of such importance that the king directed that a gold medal be bestowed upon the young author at the public sitting of the Academy in April, 1776. Two years later, at the age of thirty–five, Lavoisier was admitted a member of the Academy.

In this same year he began to devote himself almost exclusively to chemical inquiries, and established a laboratory in his home, fitted with all manner of costly apparatus and chemicals. Here he was in constant communication with the great men of science of Paris, to all of whom his doors were thrown open. One of his first undertakings in this laboratory was to demonstrate that water could not be converted into earth by repeated distillations, as was generally advocated; and to show also that there was no foundation to the existing belief that it was possible to convert water into a gas so "elastic" as to pass through the pores of a vessel. He demonstrated the fallaciousness of both these theories in 1768–1769 by elaborate experiments, a single investigation of this series occupying one hundred and one days.

In 1771 he gave the first blow to the phlogiston theory by his experiments on the calcination of metals. It will be recalled that one basis for the belief in phlogiston was the fact that when a metal was calcined it was converted into an ash, giving up its "phlogiston" in the process. To restore the metal, it was necessary to add some substance such as wheat or charcoal to the ash. Lavoisier, in examining this process of restoration, found that there was always evolved a great quantity of "air," which he supposed to be "fixed air" or carbonic acid—the same that escapes in effervescence of alkalies and calcareous earths, and in the fermentation of liquors. He then examined the process of calcination, whereby the phlogiston of the metal was supposed to have been drawn off. But far from finding that phlogiston or any other substance had been driven off, he found that something had been taken on: that the metal "absorbed air," and that the increased weight of the metal corresponded to the amount of air "absorbed." Meanwhile he was within grasp of two great discoveries, that of oxygen and of the composition of the air, which Priestley made some two years later.

The next important inquiry of this great Frenchman was as to the composition of diamonds. With the great lens of Tschirnhausen belonging to the Academy he succeeded in burning up several diamonds, regardless of expense, which, thanks to his inheritance, he could ignore. In this process he found that a gas was given off which precipitated lime from water, and proved to be carbonic acid. Observing this, and experimenting with other substances known to give off carbonic acid in the same manner, he was evidently impressed with the now well-known fact that diamond and charcoal are chemically the same. But if he did really believe it, he was cautious in expressing his belief fully. "We should never have expected," he says, "to find any relation between charcoal and diamond, and it would be unreasonable to push this analogy too far; it only exists because both substances seem to be properly ranged in the class of combustible bodies, and because they are of all these bodies the most fixed when kept from contact with air."

As we have seen, Priestley, in 1774, had discovered oxygen, or "dephlogisticated air." Four years later Lavoisier first advanced his theory that this element discovered by Priestley was the universal acidifying or oxygenating principle, which, when combined with charcoal or carbon, formed carbonic acid; when combined with sulphur, formed sulphuric (or vitriolic) acid; with nitrogen, formed nitric acid, etc., and when combined with the metals formed oxides, or calcides. Furthermore, he postulated the theory that combustion was not due to any such illusive thing as "phlogiston," since this did not exist, and it seemed to him that the phenomena of combustion heretofore attributed to phlogiston could be explained by the action of the new element oxygen and heat. This was the final blow to the phlogiston theory, which, although it had been tottering for some time, had not been completely overthrown.

In 1787 Lavoisier, in conjunction with Guyon de Morveau, Berthollet, and Fourcroy, introduced the reform in chemical nomenclature which until then had remained practically unchanged since alchemical days. Such expressions as "dephlogisticated" and "phlogisticated" would obviously have little meaning to a generation who were no longer to believe in the existence of phlogiston. It was appropriate that a revolution in chemical thought should be accompanied by a corresponding revolution in chemical names, and to Lavoisier belongs chiefly the credit of bringing about this revolution. In his Elements of Chemistry he made use of this new nomenclature, and it seemed so clearly an improvement over the old that the scientific world hastened to adopt it. In this connection Lavoisier says: "We have, therefore, laid aside the expression metallic calx altogether, and have substituted in its place the word oxide. By this it may be seen that the language we have adopted is both copious and expressive. The first or lowest degree of oxygenation in bodies converts them into oxides; a second degree of additional oxygenation constitutes the class of acids of which the specific names drawn from their particular bases terminate in ous, as in the nitrous and the sulphurous acids. The third degree of oxygenation changes these into the species of acids distinguished by the termination in ic, as the nitric and sulphuric acids; and, lastly, we can express a fourth or higher degree of oxygenation by adding the word oxygenated to the name of the acid, as has already been done with oxygenated muriatic acid."[9]

This new work when given to the world was not merely an epoch-making book; it was revolutionary. It not only discarded phlogiston altogether, but set forth that metals are simple elements, not compounds of "earth" and "phlogiston." It upheld Cavendish's demonstration that water itself, like air, is a compound of oxygen with another element. In short, it was scientific chemistry, in the modern acceptance of the term.

Lavoisier's observations on combustion are at once important and interesting: "Combustion," he says, ". . . is the decomposition of oxygen produced by a combustible body. The oxygen which forms the base of this gas is absorbed by and enters into combination with the burning body, while the caloric and light are set free. Every combustion necessarily supposes oxygenation; whereas, on the contrary, every oxygenation does not necessarily imply concomitant combustion; because combustion properly so called cannot take place without disengagement of caloric and light. Before combustion can take place, it is necessary that the base of oxygen gas should have greater affinity to the combustible body than it has to caloric; and this elective attraction, to use Bergman's expression, can only take place at a certain degree of temperature which is different for each combustible substance; hence the necessity of giving the first motion or beginning to every combustion by the approach of a heated body. This necessity of heating any body we mean to burn depends upon certain considerations which have not hitherto been attended to by any natural philosopher, for which reason I shall enlarge a little upon the subject in this place:

"Nature is at present in a state of equilibrium, which cannot have been attained until all the spontaneous combustions or oxygenations possible in an ordinary degree of temperature had taken place.... To illustrate this abstract view of the matter by example: Let us suppose the usual temperature of the earth a little changed, and it is raised only to the degree of boiling water; it is evident that in this case phosphorus, which is combustible in a considerably lower degree of temperature, would no longer exist in nature in its pure and simple state, but would always be procured in its acid or oxygenated state, and its radical would become one of the substances unknown to chemistry. By gradually increasing the temperature of the earth, the same circumstance would successively happen to all the bodies capable of combustion; and, at the last, every possible combustion having taken place, there would no longer exist any combustible body whatever, and every substance susceptible of the operation would be oxygenated and consequently incombustible.

"There cannot, therefore, exist, as far as relates to us, any combustible body but such as are non–combustible at the ordinary temperature of the earth, or, what is the same thing in other words, that it is essential to the nature of every combustible body not to possess the property of combustion unless heated, or raised to a degree of temperature at which its combustion naturally takes place. When this degree is once produced, combustion commences, and the caloric which is disengaged by the decomposition of the oxygen gas keeps up the temperature which is necessary for continuing combustion. When this is not the case—that is, when the disengaged caloric is not sufficient for keeping up the necessary temperature—the combustion ceases. This circumstance is expressed in the common language by saying that a body burns ill or with difficulty."[10]

It needed the genius of such a man as Lavoisier to complete the refutation of the false but firmly grounded phlogiston theory, and against such a book as his Elements of Chemistry the feeble weapons of the supporters of the phlogiston theory were hurled in vain.

But while chemists, as a class, had become converts to the new chemistry before the end of the century, one man, Dr. Priestley, whose work had done so much to found it, remained unconverted. In this, as in all his life–work, he showed himself to be a most remarkable man. Davy said of him, a generation later, that no other person ever discovered so many new and curious substances as he; yet to the last he was only an amateur in science, his profession, as we know, being the ministry. There is hardly another case in history of a man not a specialist in science accomplishing so much in original research as did this chemist, physiologist, electrician; the mathematician, logician, and moralist; the theologian, mental philosopher, and political economist. He took all knowledge for his field; but how he found time for his numberless researches and multifarious writings, along with his every–day duties, must ever remain a mystery to ordinary mortals.

That this marvellously receptive, flexible mind should have refused acceptance to the clearly logical doctrines of the new chemistry seems equally inexplicable. But so it was. To the very last, after all his friends had capitulated, Priestley kept up the fight. From America he sent out his last defy to the enemy, in 1800, in a brochure entitled "The Doctrine of Phlogiston Upheld," etc. In the mind of its author it was little less than a paean of victory; but all the world beside knew that it was the swan–song of the doctrine of phlogiston. Despite the defiance of this single warrior the battle was really lost and won, and as the century closed "antiphlogistic" chemistry had practical possession of the field.

III. CHEMISTRY SINCE THE TIME OF DALTON

JOHN DALTON AND THE ATOMIC THEORY

Small beginnings as have great endings—sometimes. As a case in point, note what came of the small, original effort of a self–trained back–country Quaker youth named John Dalton, who along towards the close of the eighteenth century became interested in the weather, and was led to construct and use a crude water–gauge to test the amount of the rainfall. The simple experiments thus inaugurated led to no fewer than two hundred thousand recorded observations regarding the weather, which formed the basis for some of the most epochal discoveries in meteorology, as we have seen. But this was only a beginning. The simple rain–gauge pointed the way to the most important generalization of the nineteenth century in a field of science with which, to the casual observer, it might seem to have no alliance whatever. The wonderful theory of atoms, on which the whole gigantic structure of modern chemistry is founded, was the logical outgrowth, in the mind of John Dalton, of those early studies in meteorology.

The way it happened was this: From studying the rainfall, Dalton turned naturally to the complementary process of evaporation. He was soon led to believe that vapor exists, in the atmosphere as an independent gas. But since two bodies cannot occupy the same space at the same time, this implies that the various atmospheric gases are really composed of discrete particles. These ultimate particles are so small that we cannot see them—cannot, indeed, more than vaguely imagine them—yet each particle of vapor, for example, is just as much a portion of water as if it were a drop out of the ocean, or, for that matter, the ocean itself. But, again, water is a compound substance, for it may be separated, as Cavendish has shown, into the two elementary substances hydrogen and oxygen. Hence the atom of water must be composed of two lesser atoms joined together. Imagine an atom of hydrogen and one of oxygen. Unite them, and we have an atom of water; sever them, and the water no longer exists; but whether united or separate the atoms of hydrogen and of oxygen remain hydrogen and oxygen and nothing else. Differently mixed together or united, atoms produce different gross substances; but the elementary atoms never change their chemical nature—their distinct personality.

It was about the year 1803 that Dalton first gained a full grasp of the conception of the chemical atom. At once he saw that the hypothesis, if true, furnished a marvellous key to secrets of matter hitherto insoluble—questions relating to the relative proportions of the atoms themselves. It is known, for example, that a certain bulk of hydrogen gas unites with a certain bulk of oxygen gas to form water. If it be true that this combination consists essentially of the union of atoms one with another (each single atom of hydrogen united to a single atom of oxygen), then the relative weights of the original masses of hydrogen and of oxygen must be also the relative weights of each of their respective atoms. If one pound of hydrogen unites with five and one-half pounds of oxygen (as, according to Dalton's experiments, it did), then the weight of the oxygen atom must be five and one-half times that of the hydrogen atom. Other compounds may plainly be tested in the same way. Dalton made numerous tests before he published his theory. He found that hydrogen enters into compounds in smaller proportions than any other element known to him, and so, for convenience, determined to take the weight of the hydrogen atom as unity. The atomic weight of oxygen then becomes (as given in Dalton's first table of 1803) 5.5; that of water (hydrogen plus oxygen) being of course 6.5. The atomic weights of about a score of substances are given in Dalton's first paper, which was read before the Literary and Philosophical Society of Manchester, October 21, 1803. I wonder if Dalton himself, great and acute intellect though he had, suspected, when he read that paper, that he was inaugurating one of the most fertile movements ever entered on in the whole history of science?

Be that as it may, it is certain enough that Dalton's contemporaries were at first little impressed with the novel atomic theory. Just at this time, as it chanced, a dispute was waging in the field of chemistry regarding a matter of empirical fact which must necessarily be settled before such a theory as that of Dalton could even hope for a bearing. This was the question whether or not chemical elements unite with one another always in definite proportions. Berthollet, the great co-worker with Lavoisier, and now the most authoritative of living chemists, contended that substances combine in almost indefinitely graded proportions between fixed extremes. He held that solution is really a form of chemical combination—a position which, if accepted, left no room for argument.

But this contention of the master was most actively disputed, in particular by Louis Joseph Proust, and all chemists of repute were obliged to take sides with one or the other. For a time the authority of Berthollet held out against the facts, but at last accumulated evidence told for Proust and his followers, and towards the close of the first decade of our century it came to be generally conceded that chemical elements combine with one another in fixed and definite proportions.

More than that. As the analysts were led to weigh carefully the quantities of combining elements, it was observed that the proportions are not only definite, but that they bear a very curious relation to one another. If element A combines with two different proportions of element B to form two compounds, it appears that the weight of the larger quantity of B is an exact multiple of that of the smaller quantity. This curious relation was noticed by Dr. Wollaston, one of the most accurate of observers, and a little later it was confirmed by Johan Jakob Berzelius, the great Swedish chemist, who was to be a dominating influence in the chemical world for a generation to come. But this combination of elements in numerical proportions was exactly what Dalton had noticed as early as 1802, and what bad led him directly to the atomic weights. So the confirmation of this essential point by chemists of such authority gave the strongest confirmation to the atomic theory.

During these same years the rising authority of the French chemical world, Joseph Louis Gay–Lussac, was conducting experiments with gases, which he had undertaken at first in conjunction with Humboldt, but which later on were conducted independently. In 1809, the next year after the publication of the first volume of Dalton's New System of Chemical Philosophy, Gay–Lussac published the results of his observations, and among other things brought out the remarkable fact that gases, under the same conditions as to temperature and pressure, combine always in definite numerical proportions as to volume. Exactly two volumes of hydrogen, for example, combine with one volume of oxygen to form water. Moreover, the resulting compound gas always bears a simple relation to the combining volumes. In the case just cited, the union of two volumes of hydrogen and one of oxygen results in precisely two volumes of water vapor.

Naturally enough, the champions of the atomic theory seized upon these observations of Gay-Lussac as lending strong support to their hypothesis—all of them, that is, but the curiously self-reliant and self-sufficient author of the atomic theory himself, who declined to accept the observations of the French chemist as valid. Yet the observations of Gay-Lussac were correct, as countless chemists since then have demonstrated anew, and his theory of combination by volumes became one of the foundation-stones of the atomic theory, despite the opposition of the author of that theory.

The true explanation of Gay-Lussac's law of combination by volumes was thought out almost immediately by an Italian savant, Amadeo, Avogadro, and expressed in terms of the atomic theory. The fact must be, said Avogadro, that under similar physical conditions every form of gas contains exactly the same number of ultimate particles in a given volume. Each of these ultimate physical particles may be composed of two or more atoms (as in the case of water vapor), but such a compound atom conducts itself as if it were a simple and indivisible atom, as regards the amount of space that separates it from its fellows under given conditions of pressure and temperature. The compound atom, composed of two or more elementary atoms, Avogadro proposed to distinguish, for purposes of convenience, by the name molecule. It is to the molecule, considered as the unit of physical structure, that Avogadro's law applies.

This vastly important distinction between atoms and molecules, implied in the law just expressed, was published in 1811. Four years later, the famous French physicist Ampere outlined a similar theory, and utilized the law in his mathematical calculations. And with that the law of Avogadro dropped out of sight for a full generation. Little suspecting that it was the very key to the inner mysteries of the atoms for which they were seeking, the chemists of the time cast it aside, and let it fade from the memory of their science.

This, however, was not strange, for of course the law of Avogadro is based on the atomic theory, and in 1811 the atomic theory was itself still being weighed in the balance. The law of multiple proportions found general acceptance as an empirical fact; but many of the leading lights of chemistry still looked askance at Dalton's explanation of this law. Thus Wollaston, though from the first he inclined to acceptance of the Daltonian view, cautiously suggested that it would be well to use the non-committal word "equivalent" instead of "atom"; and Davy, for a similar reason, in his book of 1812, speaks only of "proportions," binding himself to no theory as to what might be the nature of these proportions.

At least two great chemists of the time, however, adopted the atomic view with less reservation. One of these was Thomas Thomson, professor at Edinburgh, who, in 1807, had given an outline of Dalton's theory in a widely circulated book, which first brought the theory to the general attention of the chemical world. The other and even more noted advocate of the atomic theory was Johan Jakob Berzelius. This great Swedish chemist at once set to work to put the atomic theory to such tests as might be applied in the laboratory. He was an analyst of the utmost skill, and for years be devoted himself to the determination of the combining weights, "equivalents" or "proportions," of the different elements. These determinations, in so far as they were accurately made, were simple expressions of empirical facts, independent of any theory; but gradually it became more and more plain that these facts all harmonize with the atomic theory of Dalton. So by common consent the proportionate combining weights of the elements came to be known as atomic weights—the name Dalton had given them from the first—and the tangible conception of the chemical atom as a body of definite constitution and weight gained steadily in favor.

From the outset the idea had had the utmost tangibility in the mind of Dalton. He had all along represented the different atoms by geometrical symbols—as a circle for oxygen, a circle enclosing a dot for hydrogen, and the like—and had represented compounds by placing these symbols of the elements in juxtaposition. Berzelius proposed to improve upon this method by substituting for the geometrical symbol the initial of the Latin name of the element represented—O for oxygen, H for hydrogen, and so on—a numerical coefficient to follow the letter as an indication of the number of atoms present in any given compound. This simple system soon gained general acceptance, and with slight modifications it is still universally employed. Every school–boy now is aware that H_2O is the chemical way of expressing the union of two atoms of hydrogen with one of oxygen to form a molecule of water. But such a formula would have had no meaning for the wisest chemist before the day of Berzelius.

The universal fame of the great Swedish authority served to give general currency to his symbols and atomic weights, and the new point of view thus developed led presently to two important discoveries which removed the last lingering doubts as to the validity of the atomic theory. In 1819 two French physicists, Dulong and Petit, while experimenting with heat, discovered that the specific heats of solids (that is to say, the amount of heat required to raise the temperature of a given mass to a given degree) vary inversely as their atomic weights. In the same year Eilhard Mitscherlich, a German investigator, observed that compounds having the same number of atoms to the molecule are disposed to form the same angles of crystallization—a property which he called isomorphism.

Here, then, were two utterly novel and independent sets of empirical facts which harmonize strangely with the supposition that substances are composed of chemical atoms of a determinate weight. This surely could not be coincidence—it tells of law. And so as soon as the claims of Dulong and Petit and of Mitscherlich had been substantiated by other observers, the laws of the specific heat of atoms, and of isomorphism, took their place as new levers of chemical science. With the aid of these new tools an impregnable breastwork of facts was soon piled about the atomic theory. And John Dalton, the author of that theory, plain, provincial Quaker, working on to the end in semi−retirement, became known to all the world and for all time as a master of masters.

HUMPHRY DAVY AND ELECTRO−CHEMISTRY

During those early years of the nineteenth century, when Dalton was grinding away at chemical fact and theory in his obscure Manchester laboratory, another Englishman held the attention of the chemical world with a series of the most brilliant and widely heralded researches. This was Humphry Davy, a young man who had conic to London in 1801, at the instance of Count Rumford, to assume the chair of chemical philosophy in the Royal Institution, which the famous American had just founded.

Here, under Davy's direction, the largest voltaic battery yet constructed had been put in operation, and with its aid the brilliant young experimenter was expected almost to perform miracles. And indeed he scarcely disappointed the expectation, for with the aid of his battery he transformed so familiar a substance as common potash into a metal which was not only so light that it floated on water, but possessed the seemingly miraculous property of bursting into flames as soon as it came in contact with that fire−quenching liquid. If this were not a miracle, it had for the popular eye all the appearance of the miraculous.

What Davy really had done was to decompose the potash, which hitherto had been supposed to be elementary, liberating its oxygen, and thus isolating its metallic base, which he named potassium. The same thing was done with soda, and the closely similar metal sodium was discovered—metals of a unique type, possessed of a strange avidity for oxygen, and capable of seizing on it even when it is bound up in the molecules of water.

Considered as mere curiosities, these discoveries were interesting, but aside from that they were of great theoretical importance, because they showed the compound nature of some familiar chemicals that had been regarded as elements. Several other elementary earths met the same fate when subjected to the electrical influence; the metals barium, calcium, and strontium being thus discovered. Thereafter Davy always referred to the supposed elementary substances (including oxygen, hydrogen, and the rest) as "unde-compounded" bodies. These resist all present efforts to decompose them, but how can one know what might not happen were they subjected to an influence, perhaps some day to be discovered, which exceeds the battery in power as the battery exceeds the blowpipe?

Another and even more important theoretical result that flowed from Davy's experiments during this first decade of the century was the proof that no elementary substances other than hydrogen and oxygen are produced when pure water is decomposed by the electric current. It was early noticed by Davy and others that when a strong current is passed through water, alkalies appear at one pole of the battery and acids at the other, and this though the water used were absolutely pure. This seemingly told of the creation of elements—a transmutation but one step removed from the creation of matter itself—under the influence of the new "force." It was one of Davy's greatest triumphs to prove, in the series of experiments recorded in his famous Bakerian lecture of 1806, that the alleged creation of elements did not take place, the substances found at the poles of the battery having been dissolved from the walls of the vessels in which the water experimented upon had been placed. Thus the same implement which had served to give a certain philosophical warrant to the fading dreams of alchemy banished those dreams peremptorily from the domain of present science.

"As early as 1800," writes Davy, "I had found that when separate portions of distilled water, filling two glass tubes, connected by moist bladders, or any moist animal or vegetable substances, were submitted to the electrical action of the pile of Volta by means of gold wires, a nitro-muriatic solution of gold appeared in the tube containing the positive wire, or the wire transmitting the electricity, and a solution of soda in the opposite tube; but I soon ascertained that the muriatic acid owed its existence to the animal or vegetable matters employed; for when the same fibres of cotton were made use of in successive experiments, and washed after every process in a weak solution of nitric acid, the water in the apparatus containing them, though acted on for a great length of time with a very strong power, at last produced no effects upon nitrate of silver.

"In cases when I had procured much soda, the glass at its point of contact with the wire seemed considerably corroded; and I was confirmed in my idea of referring the production of the alkali principally to this source, by finding that no fixed saline matter could be obtained by electrifying distilled water in a single agate cup from two points of platina with the Voltaic battery.

"Mr. Sylvester, however, in a paper published in Mr. Nicholson's journal for last August, states that though no fixed alkali or muriatic acid appears when a single vessel is employed, yet that they are both formed when two vessels are used. And to do away with all objections with regard to vegetable substances or glass, he conducted his process in a vessel made of baked tobacco–pipe clay inserted in a crucible of platina. I have no doubt of the correctness of his results; but the conclusion appears objectionable. He conceives, that he obtained fixed alkali, because the fluid after being heated and evaporated left a matter that tinged turmeric brown, which would have happened had it been lime, a substance that exists in considerable quantities in all pipe–clay; and even allowing the presence of fixed alkali, the materials employed for the manufacture of tobacco–pipes are not at all such as to exclude the combinations of this substance.

"I resumed the inquiry; I procured small cylindrical cups of agate of the capacity of about one–quarter of a cubic inch each. They were boiled for some hours in distilled water, and a piece of very white and transparent amianthus that had been treated in the same way was made then to connect together; they were filled with distilled water and exposed by means of two platina wires to a current of electricity, from one hundred and fifty pairs of plates of copper and zinc four inches square, made active by means of solution of alum. After forty–eight hours the process was examined: Paper tinged with litmus plunged into the tube containing the transmitting or positive wire was immediately strongly reddened. Paper colored by turmeric introduced into the other tube had its color much deepened; the acid matter gave a very slight degree of turgidness to solution of nitrate of soda. The fluid that affected turmeric retained this property after being strongly boiled; and it appeared more vivid as the quantity became reduced by evaporation; carbonate of ammonia was mixed with it, and the whole dried and exposed to a strong heat; a minute quantity of white matter remained, which, as far as my examinations could go, had the properties of carbonate of soda. I compared it with similar minute portions of the pure carbonates of potash, and similar minute portions of the pure carbonates of potash and soda. It was not so deliquescent as the former of these bodies, and it formed a salt with nitric acid, which, like nitrate of soda, soon attracted moisture from a damp atmosphere and became fluid.

"This result was unexpected, but it was far from convincing me that the substances which were obtained were generated. In a similar process with glass tubes, carried on under exactly the same circumstances and for the same time, I obtained a quantity of alkali which must have been more than twenty times greater, but no traces of muriatic acid. There was much probability that the agate contained some minute portion of saline matter, not easily detected by chemical analysis, either in combination or intimate cohesion in its pores. To determine this, I repeated this a second, a third, and a fourth time. In the second experiment turbidness was still produced by a solution of nitrate of silver in the tube containing the acid, but it was less distinct; in the third process it was barely perceptible; and in the fourth process the two fluids remained perfectly clear after the mixture. The quantity of alkaline matter diminished in every operation; and in the last process, though the battery had been kept in great activity for three days, the fluid possessed, in a very slight degree, only the power of acting on paper tinged with turmeric; but its alkaline property was very sensible to litmus paper slightly reddened, which is a much more delicate test; and after evaporation and the process by carbonate of ammonia, a barely perceptible quantity of fixed alkali was still left. The acid matter in the other tube was abundant; its taste was sour; it smelled like water over which large quantities of nitrous gas have been long kept; it did not effect solution of muriate of barytes; and a drop of it placed upon a polished plate of silver left, after evaporation, a black stain, precisely similar to that produced by extremely diluted nitrous acid.

"After these results I could no longer doubt that some saline matter existing in the agate tubes had been the source of the acid matter capable of precipitating nitrate of silver and much of the alkali. Four additional repetitions of the process, however, convinced me that there was likewise some other cause for the presence of this last substance; for it continued to appear to the last in quantities sufficiently distinguishable, and apparently equal in every case. I had used every precaution, I had included the tube in glass vessels out of the reach of the circulating air; all the acting materials had been repeatedly washed with distilled water; and no part of them in contact with the fluid had been touched by the fingers.

"The only substance that I could now conceive as furnishing the fixed alkali was the water itself. This water appeared pure by the tests of nitrate of silver and muriate of barytes; but potash of soda, as is well known, rises in small quantities in rapid distillation; and the New River water which I made use of contains animal and vegetable impurities, which it was easy to conceive might furnish neutral salts capable of being carried over in vivid ebullition."[1] Further experiment proved the correctness of this inference, and the last doubt as to the origin of the puzzling chemical was dispelled.

Though the presence of the alkalies and acids in the water was explained, however, their respective migrations to the negative and positive poles of the battery remained to be accounted for. Davy's classical explanation assumed that different elements differ among themselves as to their electrical properties, some being positively, others negatively, electrified. Electricity and "chemical affinity," he said, apparently are manifestations of the same force, acting in the one case on masses, in the other on particles. Electro-positive particles unite with electro-negative particles to form chemical compounds, in virtue of the familiar principle that opposite electricities attract one another. When compounds are decomposed by the battery, this mutual attraction is overcome by the stronger attraction of the poles of the battery itself.

This theory of binary composition of all chemical compounds, through the union of electro-positive and electro-negative atoms or molecules, was extended by Berzelius, and made the basis of his famous system of theoretical chemistry. This theory held that all inorganic compounds, however complex their composition, are essentially composed of such binary combinations. For many years this view enjoyed almost undisputed sway. It received what seemed strong confirmation when Faraday showed the definite connection between the amount of electricity employed and the amount of decomposition produced in the so-called electrolyte. But its claims were really much too comprehensive, as subsequent discoveries proved.

ORGANIC CHEMISTRY AND THE IDEA OF THE MOLECULE

When Berzelius first promulgated his binary theory he was careful to restrict its unmodified application to the compounds of the inorganic world. At that time, and for a long time thereafter, it was supposed that substances of organic nature had some properties that kept them aloof from the domain of inorganic chemistry. It was little doubted that a so-called "vital force" operated here, replacing or modifying the action of ordinary "chemical affinity." It was, indeed, admitted that organic compounds are composed of familiar elements—chiefly carbon, oxygen, hydrogen, and nitrogen; but these elements were supposed to be united in ways that could not be imitated in the domain of the non-living. It was regarded almost as an axiom of chemistry that no organic compound whatever could be put together from its elements—synthesized—in the laboratory. To effect the synthesis of even the simplest organic compound, it was thought that the "vital force" must be in operation.

Therefore a veritable sensation was created in the chemical world when, in the year 1828, it was announced that the young German chemist, Friedrich Wohler, formerly pupil of Berzelius, and already known as a coming master, had actually synthesized the well-known organic product urea in his laboratory at Sacrow. The "exception which proves the rule" is something never heard of in the domain of logical science. Natural law knows no exceptions. So the synthesis of a single organic compound sufficed at a blow to break down the chemical barrier which the imagination of the fathers of the science had erected between animate and inanimate nature. Thenceforth the philosophical chemist would regard the plant and animal organisms as chemical laboratories in which conditions are peculiarly favorable for building up complex compounds of a few familiar elements, under the operation of universal chemical laws. The chimera "vital force" could no longer gain recognition in the domain of chemistry.

Now a wave of interest in organic chemistry swept over the chemical world, and soon the study of carbon compounds became as much the fashion as electrochemistry had been in the, preceding generation.

Foremost among the workers who rendered this epoch of organic chemistry memorable were Justus Liebig in Germany and Jean Baptiste Andre Dumas in France, and their respective pupils, Charles Frederic Gerhardt and Augustus Laurent. Wohler, too, must be named in the same breath, as also must Louis Pasteur, who, though somewhat younger than the others, came upon the scene in time to take chief part in the most important of the controversies that grew out of their labors.

Several years earlier than this the way had been paved for the study of organic substances by Gay-Lussac's discovery, made in 1815, that a certain compound of carbon and nitrogen, which he named cyanogen, has a peculiar degree of stability which enables it to retain its identity and enter into chemical relations after the manner of a simple body. A year later Ampere discovered that nitrogen and hydrogen, when combined in certain proportions to form what he called ammonium, have the same property. Berzelius had seized upon this discovery of the compound radical, as it was called, because it seemed to lend aid to his dualistic theory. He conceived the idea that all organic compounds are binary unions of various compound radicals with an atom of oxygen, announcing this theory in 1818. Ten years later, Liebig and Wohler undertook a joint investigation which resulted in proving that compound radicals are indeed very abundant among organic substances. Thus the theory of Berzelius seemed to be substantiated, and organic chemistry came to be defined as the chemistry of compound radicals.

But even in the day of its seeming triumph the dualistic theory was destined to receive a rude shock. This came about through the investigations of Dumas, who proved that in a certain organic substance an atom of hydrogen may be removed and an atom of chlorine substituted in its place without destroying the integrity of the original compound—much as a child might substitute one block for another in its play–house. Such a substitution would be quite consistent with the dualistic theory, were it not for the very essential fact that hydrogen is a powerfully electro–positive element, while chlorine is as strongly electro–negative. Hence the compound radical which united successively with these two elements must itself be at one time electro–positive, at another electro–negative—a seeming inconsistency which threw the entire Berzelian theory into disfavor.

In its place there was elaborated, chiefly through the efforts of Laurent and Gerhardt, a conception of the molecule as a unitary structure, built up through the aggregation of various atoms, in accordance with "elective affinities" whose nature is not yet understood A doctrine of "nuclei" and a doctrine of "types" of molecular structure were much exploited, and, like the doctrine of compound radicals, became useful as aids to memory and guides for the analyst, indicating some of the plans of molecular construction, though by no means penetrating the mysteries of chemical affinity. They are classifications rather than explanations of chemical unions. But at least they served an important purpose in giving definiteness to the idea of a molecular structure built of atoms as the basis of all substances. Now at last the word molecule came to have a distinct meaning, as distinct from "atom," in the minds of the generality of chemists, as it had had for Avogadro a third of a century before. Avogadro's hypothesis that there are equal numbers of these molecules in equal volumes of gases, under fixed conditions, was revived by Gerhardt, and a little later, under the championship of Cannizzaro, was exalted to the plane of a fixed law. Thenceforth the conception of the molecule was to be as dominant a thought in chemistry as the idea of the atom had become in a previous epoch.

CHEMICAL AFFINITY

Of course the atom itself was in no sense displaced, but Avogadro's law soon made it plain that the atom had often usurped territory that did not really belong to it. In many cases the chemists had supposed themselves dealing with atoms as units where the true unit was the molecule. In the case of elementary gases, such as hydrogen and oxygen, for example, the law of equal numbers of molecules in equal spaces made it clear that the atoms do not exist isolated, as had been supposed. Since two volumes of hydrogen unite with one volume of oxygen to form two volumes of water vapor, the simplest mathematics show, in the light of Avogadro's law, not only that each molecule of water must contain two hydrogen atoms (a point previously in dispute), but that the original molecules of hydrogen and oxygen must have been composed in each case of two atoms——else how could one volume of oxygen supply an atom for every molecule of two volumes of water?

What, then, does this imply? Why, that the elementary atom has an avidity for other atoms, a longing for companionship, an "affinity"—call it what you will—which is bound to be satisfied if other atoms are in the neighborhood. Placed solely among atoms of its own kind, the oxygen atom seizes on a fellow oxygen atom, and in all their mad dancings these two mates cling together—possibly revolving about each other in miniature planetary orbits. Precisely the same thing occurs among the hydrogen atoms. But now suppose the various pairs of oxygen atoms come near other pairs of hydrogen atoms (under proper conditions which need not detain us here), then each oxygen atom loses its attachment for its fellow, and flings itself madly into the circuit of one of the hydrogen couplets, and—presto!—there are only two molecules for every three there were before, and free oxygen and hydrogen have become water. The whole process, stated in chemical phraseology, is summed up in the statement that under the given conditions the oxygen atoms had a greater affinity for the hydrogen atoms than for one another.

As chemists studied the actions of various kinds of atoms, in regard to their unions with one another to form molecules, it gradually dawned upon them that not all elements are satisfied with the same number of companions. Some elements ask only one, and refuse to take more; while others link themselves, when occasion offers, with two, three, four, or more. Thus we saw that oxygen forsook a single atom of its own kind and linked itself with two atoms of hydrogen. Clearly, then, the oxygen atom, like a creature with two hands, is able to clutch two other atoms. But we have no proof that under any circumstances it could hold more than two. Its affinities seem satisfied when it has two bonds. But, on the other hand, the atom of nitrogen is able to hold three atoms of hydrogen, and does so in the molecule of ammonium (NH_3); while the carbon atom can hold four atoms of hydrogen or two atoms of oxygen.

Evidently, then, one atom is not always equivalent to another atom of a different kind in combining powers. A recognition of this fact by Frankland about 1852, and its further investigation by others (notably A. Kekule and A. S. Couper), led to the introduction of the word equivalent into chemical terminology in a new sense, and in particular to an understanding of the affinities or "valency" of different elements, which proved of the most fundamental importance. Thus it was shown that, of the four elements that enter most prominently into organic compounds, hydrogen can link itself with only a single bond to any other element—it has, so to speak, but a single hand with which to grasp—while oxygen has capacity for two bonds, nitrogen for three (possibly for five), and carbon for four. The words monovalent, divalent, trivalent, tretrava–lent, etc., were coined to express this most important fact, and the various elements came to be known as monads, diads, triads, etc. Just why different elements should differ thus in valency no one as yet knows; it is an empirical fact that they do. And once the nature of any element has been determined as regards its valency, a most important insight into the possible behavior of that element has been secured. Thus a consideration of the fact that hydrogen is monovalent, while oxygen is divalent, makes it plain that we must expect to find no more than three compounds of these two elements—namely, H—O—(written HO by the chemist, and called hydroxyl); H—O—H (H_2O, or water), and H—O—O—H (H_2O_2, or hydrogen peroxide). It will be observed that in the first of these compounds the atom of oxygen stands, so to speak, with one of its hands free, eagerly reaching out, therefore, for another companion, and hence, in the language of chemistry, forming an unstable compound. Again, in the third compound, though all hands are clasped, yet one pair links oxygen with oxygen; and this also must be an unstable union, since the avidity of an atom for its own kind is relatively weak. Thus the well–known properties of hydrogen peroxide are explained, its easy decomposition, and the eagerness with which it seizes upon the elements of other compounds.

But the molecule of water, on the other hand, has its atoms arranged in a state of stable equilibrium, all their affinities being satisfied. Each hydrogen atom has satisfied its own affinity by clutching the oxygen atom; and the oxygen atom has both its bonds satisfied by clutching back at the two hydrogen atoms. Therefore the trio, linked in this close bond, have no tendency to reach out for any other companion, nor, indeed, any power to hold another should it thrust itself upon them. They form a "stable" compound, which under all ordinary circumstances will retain its identity as a molecule of water, even though the physical mass of which it is a part changes its condition from a solid to a gas from ice to vapor.

But a consideration of this condition of stable equilibrium in the molecule at once suggests a new question: How can an aggregation of atoms, having all their affinities satisfied, take any further part in chemical reactions? Seemingly such a molecule, whatever its physical properties, must be chemically inert, incapable of any atomic readjustments. And so in point of fact it is, so long as its component atoms cling to one another unremittingly. But this, it appears, is precisely what the atoms are little prone to do. It seems that they are fickle to the last degree in their individual attachments, and are as prone to break away from bondage as they are to enter into it. Thus the oxygen atom which has just flung itself into the circuit of two hydrogen atoms, the next moment flings itself free again and seeks new companions. It is for all the world like the incessant change of partners in a rollicking dance. This incessant dissolution and reformation of molecules in a substance which as a whole remains apparently unchanged was first fully appreciated by Ste.–Claire Deville, and by him named dissociation. It is a process which goes on much more actively in some compounds than in others, and very much more actively under some physical conditions (such as increase of temperature) than under others. But apparently no substances at ordinary temperatures, and no temperature above the absolute zero, are absolutely free from its disturbing influence. Hence it is that molecules having all the valency of their atoms fully satisfied do not lose their chemical activity—since each atom is momentarily free in the exchange of partners, and may seize upon different atoms from its former partners, if those it prefers are at hand.

While, however, an appreciation of this ceaseless activity of the atom is essential to a proper understanding of its chemical efficiency, yet from another point of view the "saturated" molecule—that is, the molecule whose atoms have their valency all satisfied—may be thought of as a relatively fixed or stable organism. Even though it may presently be torn down, it is for the time being a completed structure; and a consideration of the valency of its atoms gives the best clew that has hitherto been obtainable as to the character of its architecture. How important this matter of architecture of the molecule—of space relations of the atoms—may be was demonstrated as long ago as 1823, when Liebig and Wohler proved, to the utter bewilderment of the chemical world, that two substances may have precisely the same chemical constitution—the same number and kind of atoms—and yet differ utterly in physical properties. The word isomerism was coined by Berzelius to express this anomalous condition of things, which seemed to negative the most fundamental truths of chemistry. Naming the condition by no means explained it, but the fact was made clear that something besides the mere number and kind of atoms is important in the architecture of a molecule. It became certain that atoms are not thrown together haphazard to build a molecule, any more than bricks are thrown together at random to form a house.

How delicate may be the gradations of architectural design in building a molecule was well illustrated about 1850, when Pasteur discovered that some carbon compounds—as certain sugars—can only be distinguished from one another, when in solution, by the fact of their twisting or polarizing a ray of light to the left or to the right, respectively. But no inkling of an explanation of these strange variations of molecular structure came until the discovery of the law of valency. Then much of the mystery was cleared away; for it was plain that since each atom in a molecule can hold to itself only a fixed number of other atoms, complex molecules must have their atoms linked in definite chains or groups. And it is equally plain that where the atoms are numerous, the exact plan of grouping may sometimes be susceptible of change without doing violence to the law of valency. It is in such cases that isomerism is observed to occur.

By paying constant heed to this matter of the affinities, chemists are able to make diagrammatic pictures of the plan of architecture of any molecule whose composition is known. In the simple molecule of water (H2O), for example, the two hydrogen atoms must have released each other before they could join the oxygen, and the manner of linking must apparently be that represented in the graphic formula H—O—H. With molecules composed of a large number of atoms, such graphic representation of the scheme of linking is of course increasingly difficult, yet, with the affinities for a guide, it is always possible. Of course no one supposes that such a formula, written in a single plane, can possibly represent the true architecture of the molecule: it is at best suggestive or diagrammatic rather than pictorial. Nevertheless, it affords hints as to the structure of the molecule such as the fathers of chemistry would not have thought it possible ever to attain.

PERIODICITY OF ATOMIC WEIGHTS

These utterly novel studies of molecular architecture may seem at first sight to take from the atom much of its former prestige as the all–important personage of the chemical world. Since so much depends upon the mere position of the atoms, it may appear that comparatively little depends upon the nature of the atoms themselves. But such a view is incorrect, for on closer consideration it will appear that at no time has the atom been seen to renounce its peculiar personality. Within certain limits the character of a molecule may be altered by changing the positions of its atoms (just as different buildings may be constructed of the same bricks), but these limits are sharply defined, and it would be as impossible to exceed them as it would be to build a stone building with bricks. From first to last the brick remains a brick, whatever the style of architecture it helps to construct; it never becomes a stone. And just as closely does each atom retain its own peculiar properties, regardless of its surroundings.

Thus, for example, the carbon atom may take part in the formation at one time of a diamond, again of a piece of coal, and yet again of a particle of sugar, of wood fibre, of animal tissue, or of a gas in the atmosphere; but from first to last—from glass–cutting gem to intangible gas—there is no demonstrable change whatever in any single property of the atom itself. So far as we know, its size, its weight, its capacity for vibration or rotation, and its inherent affinities, remain absolutely unchanged throughout all these varying fortunes of position and association. And the same thing is true of every atom of all of the seventy–odd elementary substances with which the modern chemist is acquainted. Every one appears always to maintain its unique integrity, gaining nothing and losing nothing.

All this being true, it would seem as if the position of the Daltonian atom as a primordial bit of matter, indestructible and non–transmutable, had been put to the test by the chemistry of our century, and not found wanting. Since those early days of the century when the electric battery performed its miracles and seemingly reached its limitations in the hands of Davy, many new elementary substances have been discovered, but no single element has been displaced from its position as an undecomposable body. Rather have the analyses of the chemist seemed to make it more and more certain that all elementary atoms are in truth what John Herschel called them, "manufactured articles"—primordial, changeless, indestructible.

And yet, oddly enough, it has chanced that hand in hand with the experiments leading to such a goal have gone other experiments arid speculations of exactly the opposite tenor. In each generation there have been chemists among the leaders of their science who have refused to admit that the so–called elements are really elements at all in any final sense, and who have sought eagerly for proof which might warrant their scepticism. The first bit of evidence tending to support this view was furnished by an English physician, Dr. William Prout, who in 1815 called attention to a curious relation to be observed between the atomic weight of the various elements. Accepting the figures given by the authorities of the time (notably Thomson and Berzelius), it appeared that a strikingly large proportion of the atomic weights were exact multiples of the weight of hydrogen, and that others differed so slightly that errors of observation might explain the discrepancy. Prout felt that it could not be accidental, and he could think of no tenable explanation, unless it be that the atoms of the various alleged elements are made up of different fixed numbers of hydrogen atoms. Could it be that the one true element—the one primal matter—is hydrogen, and that all other forms of matter are but compounds of this original substance?

Prout advanced this startling idea at first tentatively, in an anonymous publication; but afterwards he espoused it openly and urged its tenability. Coming just after Davy's dissociation of some supposed elements, the idea proved alluring, and for a time gained such popularity that chemists were disposed to round out the observed atomic weights of all elements into whole numbers. But presently renewed determinations of the atomic weights seemed to discountenance this practice, and Prout's alleged law fell into disrepute. It was revived, however, about 1840, by Dumas, whose great authority secured it a respectful hearing, and whose careful redetermination of the weight of carbon, making it exactly twelve times that of hydrogen, aided the cause.

Subsequently Stas, the pupil of Dumas, undertook a long series of determinations of atomic weights, with the expectation of confirming the Proutian hypothesis. But his results seemed to disprove the hypothesis, for the atomic weights of many elements differed from whole numbers by more, it was thought, than the limits of error of the experiments. It was noteworthy, however, that the confidence of Dumas was not shaken, though he was led to modify the hypothesis, and, in accordance with previous suggestions of Clark and of Marignac, to recognize as the primordial element, not hydrogen itself, but an atom half the weight, or even one–fourth the weight, of that of hydrogen, of which primordial atom the hydrogen atom itself is compounded. But even in this modified form the hypothesis found great opposition from experimental observers.

In 1864, however, a novel relation between the weights of the elements and their other characteristics was called to the attention of chemists by Professor John A. R. Newlands, of London, who had noticed that if the elements are arranged serially in the numerical order of their atomic weights, there is a curious recurrence of similar properties at intervals of eight elements This so-called "law of octaves" attracted little immediate attention, but the facts it connotes soon came under the observation of other chemists, notably of Professors Gustav Hinrichs in America, Dmitri Mendeleeff in Russia, and Lothar Meyer in Germany. Mendeleeff gave the discovery fullest expression, explicating it in 1869, under the title of "the periodic law."

Though this early exposition of what has since been admitted to be a most important discovery was very fully outlined, the generality of chemists gave it little heed till a decade or so later, when three new elements, gallium, scandium, and germanium, were discovered, which, on being analyzed, were quite unexpectedly found to fit into three gaps which Mendeleeff had left in his periodic scale. In effect the periodic law had enabled Mendeleeff to predicate the existence of the new elements years before they were discovered. Surely a system that leads to such results is no mere vagary. So very soon the periodic law took its place as one of the most important generalizations of chemical science.

This law of periodicity was put forward as an expression of observed relations independent of hypothesis; but of course the theoretical bearings of these facts could not be overlooked. As Professor J. H. Gladstone has said, it forces upon us "the conviction that the elements are not separate bodies created without reference to one another, but that they have been originally fashioned, or have been built up, from one another, according to some general plan." It is but a short step from that proposition to the Proutian hypothesis.

NEW WEAPONS—SPECTROSCOPE AND CAMERA

But the atomic weights are not alone in suggesting the compound nature of the alleged elements. Evidence of a totally different kind has contributed to the same end, from a source that could hardly have been imagined when the Proutian hypothesis, was formulated, through the tradition of a novel weapon to the armamentarium of the chemist—the spectroscope. The perfection of this instrument, in the hands of two German scientists, Gustav Robert Kirchhoff and Robert Wilhelm Bunsen, came about through the investigation, towards the middle of the century, of the meaning of the dark lines which had been observed in the solar spectrum by Fraunhofer as early as 1815, and by Wollaston a decade earlier. It was suspected by Stokes and by Fox Talbot in England, but first brought to demonstration by Kirchhoff and Bunsen, that these lines, which were known to occupy definite positions in the spectrum, are really indicative of particular elementary substances. By means of the spectroscope, which is essentially a magnifying lens attached to a prism of glass, it is possible to locate the lines with great accuracy, and it was soon shown that here was a new means of chemical analysis of the most exquisite delicacy. It was found, for example, that the spectroscope could detect the presence of a quantity of sodium so infinitesimal as the one two–hundred–thousandth of a grain. But what was even more important, the spectroscope put no limit upon the distance of location of the substance it tested, provided only that sufficient light came from it. The experiments it recorded might be performed in the sun, or in the most distant stars or nebulae; indeed, one of the earliest feats of the instrument was to wrench from the sun the secret of his chemical constitution.

To render the utility of the spectroscope complete, however, it was necessary to link with it another new chemical agency—namely, photography. This now familiar process is based on the property of light to decompose certain unstable compounds of silver, and thus alter their chemical composition. Davy and Wedgwood barely escaped the discovery of the value of the photographic method early in the nineteenth century. Their successors quite overlooked it until about 1826, when Louis J. M. Daguerre, the French chemist, took the matter in hand, and after many years of experimentation brought it to relative perfection in 1839, in which year the famous daguerreotype first brought the matter to popular attention. In the same year Mr. Fox Talbot read a paper on the subject before the Royal Society, and soon afterwards the efforts of Herschel and numerous other natural philosophers contributed to the advancement of the new method.

In 1843 Dr. John W. Draper, the famous English–American chemist and physiologist, showed that by photography the Fraunhofer lines in the solar spectrum might be mapped with absolute accuracy; also proving that the silvered film revealed many lines invisible to the unaided eye. The value of this method of observation was recognized at once, and, as soon as the spectroscope was perfected, the photographic method, in conjunction with its use, became invaluable to the chemist. By this means comparisons of spectra may be made with a degree of accuracy not otherwise obtainable; and, in case of the stars, whole clusters of spectra may be placed on record at a single observation.

As the examination of the sun and stars proceeded, chemists were amazed or delighted, according to their various preconceptions, to witness the proof that many familiar terrestrial elements are to be found in the celestial bodies. But what perhaps surprised them most was to observe the enormous preponderance in the sidereal bodies of the element hydrogen. Not only are there vast quantities of this element in the sun's atmosphere, but some other suns appeared to show hydrogen lines almost exclusively in their spectra. Presently it appeared that the stars of which this is true are those white stars, such as Sirius, which had been conjectured to be the hottest; whereas stars that are only red–hot, like our sun, show also the vapors of many other elements, including iron and other metals.

In 1878 Professor J. Norman Lockyer, in a paper before the Royal Society, called attention to the possible significance of this series of observations. He urged that the fact of the sun showing fewer elements than are observed here on the cool earth, while stars much hotter than the sun show chiefly one element, and that one hydrogen, the lightest of known elements, seemed to give color to the possibility that our alleged elements are really compounds, which at the temperature of the hottest stars may be decomposed into hydrogen, the latter "element" itself being also doubtless a compound, which might be resolved under yet more trying conditions.

Here, then, was what might be termed direct experimental evidence for the hypothesis of Prout. Unfortunately, however, it is evidence of a kind which only a few experts are competent to discuss—so very delicate a matter is the spectral analysis of the stars. What is still more unfortunate, the experts do not agree among themselves as to the validity of Professor Lockyer's conclusions. Some, like Professor Crookes, have accepted them with acclaim, hailing Lockyer as "the Darwin of the inorganic world," while others have sought a different explanation of the facts he brings forward. As yet it cannot be said that the controversy has been brought to final settlement. Still, it is hardly to be doubted that now, since the periodic law has seemed to join hands with the spectroscope, a belief in the compound nature of the so–called elements is rapidly gaining ground among chemists. More and more general becomes the belief that the Daltonian atom is really a compound radical, and that back of the seeming diversity of the alleged elements is a single form of primordial matter. Indeed, in very recent months, direct experimental evidence for this view has at last come to hand, through the study of radio–active substances. In a later chapter we shall have occasion to inquire how this came about.

IV. ANATOMY AND PHYSIOLOGY IN THE EIGHTEENTH CENTURY

ALBRECHT VON HALLER

An epoch in physiology was made in the eighteenth century by the genius and efforts of Albrecht von Haller (1708–1777), of Berne, who is perhaps as worthy of the title "The Great" as any philosopher who has been so christened by his contemporaries since the time of Hippocrates. Celebrated as a physician, he was proficient in various fields, being equally famed in his own time as poet, botanist, and statesman, and dividing his attention between art and science.

As a child Haller was so sickly that he was unable to amuse himself with the sports and games common to boys of his age, and so passed most of his time poring over books. When ten years of age he began writing poems in Latin and German, and at fifteen entered the University of Tubingen. At seventeen he wrote learned articles in opposition to certain accepted doctrines, and at nineteen he received his degree of doctor. Soon after this he visited England, where his zeal in dissecting brought him under suspicion of grave–robbery, which suspicion made it expedient for him to return to the Continent. After studying botany in Basel for some time he made an extended botanical journey through Switzerland, finally settling in his native city, Berne, as a practising physician. During this time he did not neglect either poetry or botany, publishing anonymously a collection of poems.

In 1736 he was called to Gottingen as professor of anatomy, surgery, chemistry, and botany. During his labors in the university he never neglected his literary work, sometimes living and sleeping for days and nights together in his library, eating his meals while delving in his books, and sleeping only when actually compelled to do so by fatigue. During all this time he was in correspondence with savants from all over the world, and it is said of him that he never left a letter of any kind unanswered.

Haller's greatest contribution to medical science was his famous doctrine of irritability, which has given him the name of "father of modern nervous physiology," just as Harvey is called "the father of the modern physiology of the blood." It has been said of this famous doctrine of irritability that "it moved all the minds of the century—and not in the departments of medicine alone—in a way of which we of the present day have no satisfactory conception, unless we compare it with our modern Darwinism."[1]

The principle of general irritability had been laid down by Francis Glisson (1597–1677) from deductive studies, but Haller proved by experiments along the line of inductive methods that this irritability was not common to all "fibre as well as to the fluids of the body," but something entirely special, and peculiar only to muscular substance. He distinguished between irritability of muscles and sensibility of nerves. In 1747 he gave as the three forces that produce muscular movements: elasticity, or "dead nervous force"; irritability, or "innate nervous force"; and nervous force in itself. And in 1752 he described one hundred and ninety experiments for determining what parts of the body possess "irritability"—that is, the property of contracting when stimulated. His conclusion that this irritability exists in muscular substance alone and is quite independent of the nerves proceeding to it aroused a controversy that was never definitely settled until late in the nineteenth century, when Haller's theory was found to be entirely correct.

It was in pursuit of experiments to establish his theory of irritability that Haller made his chief discoveries in embryology and development. He proved that in the process of incubation of the egg the first trace of the heart of the chick shows itself in the thirty–eighth hour, and that the first trace of red blood showed in the forty–first hour. By his investigations upon the lower animals he attempted to confirm the theory that since the creation of genus every individual is derived from a preceding individual—the existing theory of preformation, in which he believed, and which taught that "every individual is fully and completely preformed in the germ, simply growing from microscopic to visible proportions, without developing any new parts."

In physiology, besides his studies of the nervous system, Haller studied the mechanism of respiration, refuting the teachings of Hamberger (1697–1755), who maintained that the lungs contract independently. Haller, however, in common with his contemporaries, failed utterly to understand the true function of the lungs. The great physiologist's influence upon practical medicine, while most profound, was largely indirect. He was a theoretical rather than a practical physician, yet he is credited with being the first physician to use the watch in counting the pulse.

BATTISTA MORGAGNI AND MORBID ANATOMY

A great contemporary of Haller was Giovanni Battista Morgagni (1682–1771), who pursued what Sydenham had neglected, the investigation in anatomy, thus supplying a necessary counterpart to the great Englishman's work. Morgagni's investigations were directed chiefly to the study of morbid anatomy—the study of the structure of diseased tissue, both during life and post mortem, in contrast to the normal anatomical structures. This work cannot be said to have originated with him; for as early as 1679 Bonnet had made similar, although less extensive, studies; and later many investigators, such as Lancisi and Haller, had made post–mortem studies. But Morgagni's De sedibus et causis morborum per anatomen indagatis was the largest, most accurate, and best–illustrated collection of cases that had ever been brought together, and marks an epoch in medical science. From the time of the publication of Morgagni's researches, morbid anatomy became a recognized branch of the medical science, and the effect of the impetus thus given it has been steadily increasing since that time.

WILLIAM HUNTER

William Hunter (1718–1783) must always be remembered as one of the greatest physicians and anatomists of the eighteenth century, and particularly as the first great teacher of anatomy in England; but his fame has been somewhat overshadowed by that of his younger brother John.

Hunter had been intended and educated for the Church, but on the advice of the surgeon William Cullen he turned his attention to the study of medicine. His first attempt at teaching was in 1746, when he delivered a series of lectures on surgery for the Society of Naval Practitioners. These lectures proved so interesting and instructive that he was at once invited to give others, and his reputation as a lecturer was soon established. He was a natural orator and story–teller, and he combined with these attractive qualities that of thoroughness and clearness in demonstrations, and although his lectures were two hours long he made them so full of interest that his pupils seldom tired of listening. He believed that he could do greater good to the world by "publicly teaching his art than by practising it," and even during the last few days of his life, when he was so weak that his friends remonstrated against it, he continued his teaching, fainting from exhaustion at the end of his last lecture, which preceded his death by only a few days.

For many years it was Hunter's ambition to establish a museum where the study of anatomy, surgery, and medicine might be advanced, and in 1765 he asked for a grant of a plot of ground for this purpose, offering to spend seven thousand pounds on its, erection besides endowing it with a professorship of anatomy. Not being able to obtain this grant, however, he built a house, in which were lecture and dissecting rooms, and his museum. In this museum were anatomical preparations, coins, minerals, and natural—history specimens.

Hunter's weakness was his love of controversy and his resentment of contradiction. This brought him into strained relations with many of the leading physicians of his time, notably his own brother John, who himself was probably not entirely free from blame in the matter. Hunter is said to have excused his own irritability on the grounds that being an anatomist, and accustomed to "the passive submission of dead bodies," contradictions became the more unbearable. Many of the physiological researches begun by him were carried on and perfected by his more famous brother, particularly his investigations of the capillaries, but he added much to the anatomical knowledge of several structures of the body, notably as to the structure of cartilages and joints.

JOHN HUNTER

In Abbot Islip's chapel in Westminster Abbey, close to the resting—place of Ben Jonson, rest the remains of John Hunter (1728–1793), famous in the annals of medicine as among the greatest physiologists and surgeons that the world has ever produced: a man whose discoveries and inventions are counted by scores, and whose field of research was only limited by the outermost boundaries of eighteenth—century science, although his efforts were directed chiefly along the lines of his profession.

Until about twenty years of age young Hunter had shown little aptitude for study, being unusually fond of out—door sports and amusements; but about that time, realizing that some occupation must be selected, he asked permission of his brother William to attempt some dissections in his anatomical school in London. To the surprise of his brother he made this dissection unusually well; and being given a second, he acquitted himself with such skill that his brother at once predicted that he would become a great anatomist. Up to this time he had had no training of any kind to prepare him for his professional career, and knew little of Greek or Latin—languages entirely unnecessary for him, as he proved in all of his life work. Ottley tells the story that, when twitted with this lack of knowledge of the "dead languages" in after life, he said of his opponent, "I could teach him that on the dead body which he never knew in any language, dead or living."

By his second year in dissection he had become so skilful that he was given charge of some of the classes in his brother's school; in 1754 he became a surgeon's pupil in St. George's Hospital, and two years later house–surgeon. Having by overwork brought on symptoms that seemed to threaten consumption, he accepted the position of staff–surgeon to an expedition to Belleisle in 1760, and two years later was serving with the English army at Portugal. During all this time he was constantly engaged in scientific researches, many of which, such as his observations of gun–shot wounds, he put to excellent use in later life. On returning to England much improved in health in 1763, he entered at once upon his career as a London surgeon, and from that time forward his progress was a practically uninterrupted series of successes in his profession.

Hunter's work on the study of the lymphatics was of great service to the medical profession. This important net–work of minute vessels distributed throughout the body had recently been made the object of much study, and various students, including Haller, had made extensive investigations since their discovery by Asellius. But Hunter, in 1758, was the first to discover the lymphatics in the neck of birds, although it was his brother William who advanced the theory that the function of these vessels was that of absorbents. One of John Hunter's pupils, William Hewson (1739–1774), first gave an account, in 1768, of the lymphatics in reptiles and fishes, and added to his teacher's investigations of the lymphatics in birds. These studies of the lymphatics have been regarded, perhaps with justice, as Hunter's most valuable contributions to practical medicine.

In 1767 he met with an accident by which he suffered a rupture of the tendo Achillis—the large tendon that forms the attachment of the muscles of the calf to the heel. From observations of this accident, and subsequent experiments upon dogs, he laid the foundation for the now simple and effective operation for the cure of club feet and other deformities involving the tendons. In 1772 he moved into his residence at Earlscourt, Brompton, where he gathered about him a great menagerie of animals, birds, reptiles, insects, and fishes, which he used in his physiological and surgical experiments. Here he performed a countless number of experiments—more, probably, than "any man engaged in professional practice has ever conducted." These experiments varied in nature from observations of the habits of bees and wasps to major surgical operations performed upon hedgehogs, dogs, leopards, etc. It is said that for fifteen years he kept a flock of geese for the sole purpose of studying the process of development in eggs.

Hunter began his first course of lectures in 1772, being forced to do this because he had been so repeatedly misquoted, and because he felt that he could better gauge his own knowledge in this way. Lecturing was a sore trial to him, as he was extremely diffident, and without writing out his lectures in advance he was scarcely able to speak at all. In this he presented a marked contrast to his brother William, who was a fluent and brilliant speaker. Hunter's lectures were at best simple readings of the facts as he had written them, the diffident teacher seldom raising his eyes from his manuscript and rarely stopping until his complete lecture had been read through. His lectures were, therefore, instructive rather than interesting, as he used infinite care in preparing them; but appearing before his classes was so dreaded by him that he is said to have been in the habit of taking a half—drachm of laudanum before each lecture to nerve him for the ordeal. One is led to wonder by what name he shall designate that quality of mind that renders a bold and fearless surgeon like Hunter, who is undaunted in the face of hazardous and dangerous operations, a stumbling, halting, and "frightened" speaker before a little band of, at most, thirty young medical students. And yet this same thing is not unfrequently seen among the boldest surgeons.

Hunter's Operation for the Cure of Aneurisms

It should be an object—lesson to those who, ignorantly or otherwise, preach against the painless vivisection as practised to—day, that by the sacrifice of a single deer in the cause of science Hunter discovered a fact in physiology that has been the means of saving thousands of human lives and thousands of human bodies from needless mutilation. We refer to the discovery of the "collateral circulation" of the blood, which led, among other things, to Hunter's successful operation upon aneurisms.

Simply stated, every organ or muscle of the body is supplied by one large artery, whose main trunk distributes the blood into its lesser branches, and thence through the capillaries. Cutting off this main artery, it would seem, should cut off entirely the blood—supply to the particular organ which is supplied by this vessel; and until the time of Hunter's demonstration this belief was held by most physiologists. But nature has made a provision for this possible stoppage of blood—supply from a single source, and has so arranged that some of the small arterial branches coming from the main supply—trunk are connected with other arterial branches coming from some other supply—trunk. Under normal conditions the main arterial trunks supply their respective organs, the little connecting arterioles playing an insignificant part. But let the main supply–trunk be cut off or stopped for whatever reason, and a remarkable thing takes place. The little connecting branches begin at once to enlarge and draw blood from the neighboring uninjured supply–trunk, This enlargement continues until at last a new route for the circulation has been established, the organ no longer depending on the now defunct original arterial trunk, but getting on as well as before by this "collateral" circulation that has been established.

The thorough understanding of this collateral circulation is one of the most important steps in surgery, for until it was discovered amputations were thought necessary in such cases as those involving the artery supplying a leg or arm, since it was supposed that, the artery being stopped, death of the limb and the subsequent necessity for amputation were sure to follow. Hunter solved this problem by a single operation upon a deer, and his practicality as a surgeon led him soon after to apply this knowledge to a certain class of surgical cases in a most revolutionary and satisfactory manner.

What led to Hunter's far–reaching discovery was his investigation as to the cause of the growth of the antlers of the deer. Wishing to ascertain just what part the blood–supply on the opposite sides of the neck played in the process of development, or, perhaps more correctly, to see what effect cutting off the main blood–supply would have, Hunter had one of the deer of Richmond Park caught and tied, while he placed a ligature around one of the carotid arteries—one of the two principal arteries that supply the head with blood. He observed that shortly after this the antler (which was only half grown and consequently very vascular) on the side of the obliterated artery became cold to the touch—from the lack of warmth–giving blood. There was nothing unexpected in this, and Hunter thought nothing of it until a few days later, when he found, to his surprise, that the antler had become as warm as its fellow, and was apparently increasing in size. Puzzled as to how this could be, and suspecting that in some way his ligature around the artery had not been effective, he ordered the deer killed, and on examination was astonished to find that while his ligature had completely shut off the blood–supply from the source of that carotid artery, the smaller arteries had become enlarged so as to supply the antler with blood as well as ever, only by a different route.

Hunter soon had a chance to make a practical application of the knowledge thus acquired. This was a case of popliteal aneurism, operations for which had heretofore proved pretty uniformly fatal. An aneurism, as is generally understood, is an enlargement of a certain part of an artery, this enlargement sometimes becoming of enormous size, full of palpitating blood, and likely to rupture with fatal results at any time. If by any means the blood can be allowed to remain quiet for even a few hours in this aneurism it will form a clot, contract, and finally be absorbed and disappear without any evil results. The problem of keeping the blood quiet, with the heart continually driving it through the vessel, is not a simple one, and in Hunter's time was considered so insurmountable that some surgeons advocated amputation of any member having an aneurism, while others cut down upon the tumor itself and attempted to tie off the artery above and below. The first of these operations maimed the patient for life, while the second was likely to prove fatal.

In pondering over what he had learned about collateral circulation and the time required for it to become fully established, Hunter conceived the idea that if the blood–supply was cut off from above the aneurism, thus temporarily preventing the ceaseless pulsations from the heart, this blood would coagulate and form a clot before the collateral circulation could become established or could affect it. The patient upon whom he performed his now celebrated operation was afflicted with a popliteal aneurism—that is, the aneurism was located on the large popliteal artery just behind the knee–joint. Hunter, therefore, tied off the femoral, or main supplying artery in the thigh, a little distance above the aneurism. The operation was entirely successful, and in six weeks' time the patient was able to leave the hospital, and with two sound limbs. Naturally the simplicity and success of this operation aroused the attention of Europe, and, alone, would have made the name of Hunter immortal in the annals of surgery. The operation has ever since been called the "Hunterian" operation for aneurism, but there is reason to believe that Dominique Anel (born about 1679) performed a somewhat similar operation several years earlier. It is probable, however, that Hunter had never heard of this work of Anel, and that his operation was the outcome of his own independent reasoning from the facts he had learned about collateral circulation. Furthermore, Hunter's mode of operation was a much better one than Anel's, and, while Anel's must claim priority, the credit of making it widely known will always be Hunter's.

The great services of Hunter were recognized both at home and abroad, and honors and positions of honor and responsibility were given him. In 1776 he was appointed surgeon–extraordinary to the king; in 1783 he was elected a member of the Royal Society of Medicine and of the Royal Academy of Surgery at Paris; in 1786 he became deputy surgeon–general of the army; and in 1790 he was appointed surgeon–general and inspector–general of hospitals. All these positions he filled with credit, and he was actively engaged in his tireless pursuit of knowledge and in discharging his many duties when in October, 1793, he was stricken while addressing some colleagues, and fell dead in the arms of a fellow–physician.

LAZZARO SPALLANZANI

Hunter's great rival among contemporary physiologists was the Italian Lazzaro Spallanzani (1729–1799), one of the most picturesque figures in the history of science. He was not educated either as a scientist or physician, devoting, himself at first to philosophy and the languages, afterwards studying law, and later taking orders. But he was a keen observer of nature and of a questioning and investigating mind, so that he is remembered now chiefly for his discoveries and investigations in the biological sciences. One important demonstration was his controversion of the theory of abiogenesis, or "spontaneous generation," as propounded by Needham and Buffon. At the time of Needham's experiments it had long been observed that when animal or vegetable matter had lain in water for a little time—long enough for it to begin to undergo decomposition—the water became filled with microscopic creatures, the "infusoria animalculis." This would tend to show, either that the water or the animal or vegetable substance contained the "germs" of these minute organisms, or else that they were generated spontaneously. It was known that boiling killed these animalcules, and Needham agreed, therefore, that if he first heated the meat or vegetables, and also the water containing them, and then placed them in hermetically sealed jars—if he did this, and still the animalcules made their appearance, it would be proof–positive that they had been generated spontaneously. Accordingly be made numerous experiments, always with the same results—that after a few days the water was found to swarm with the microscopic creatures. The thing seemed proven beyond question—providing, of course, that there had been no slips in the experiments.

But Abbe Spallanzani thought that he detected such slips in Needham's experiment. The possibility of such slips might come in several ways: the contents of the jar might not have been boiled for a sufficient length of time to kill all the germs, or the air might not have been excluded completely by the sealing process. To cover both these contingencies, Spallanzani first hermetically sealed the glass vessels and then boiled them for three–quarters of an hour. Under these circumstances no animalcules ever made their appearance—a conclusive demonstration that rendered Needham's grounds for his theory at once untenable.[2]

Allied to these studies of spontaneous generation were Spallanzani's experiments and observations on the physiological processes of generation among higher animals. He experimented with frogs, tortoises, and dogs; and settled beyond question the function of the ovum and spermatozoon. Unfortunately he misinterpreted the part played by the spermatozoa in believing that their surrounding fluid was equally active in the fertilizing process, and it was not until some forty years later (1824) that Dumas corrected this error.

THE CHEMICAL THEORY OF DIGESTION

Among the most interesting researches of Spallanzani were his experiments to prove that digestion, as carried on in the stomach, is a chemical process. In this he demonstrated, as Rene Reaumur had attempted to demonstrate, that digestion could be carried on outside the walls of the stomach as an ordinary chemical reaction, using the gastric juice as the reagent for performing the experiment. The question as to whether the stomach acted as a grinding or triturating organ, rather than as a receptacle for chemical action, had been settled by Reaumur and was no longer a question of general dispute. Reaumur had demonstrated conclusively that digestion would take place in the stomach in the same manner and the same time if the substance to be digested was protected from the peristalic movements of the stomach and subjected to the action of the gastric juice only. He did this by introducing the substances to be digested into the stomach in tubes, and thus protected so that while the juices of the stomach could act upon them freely they would not be affected by any movements of the organ.

Following up these experiments, he attempted to show that digestion could take place outside the body as well as in it, as it certainly should if it were a purely chemical process. He collected quantities of gastric juice, and placing it in suitable vessels containing crushed grain or flesh, kept the mixture at about the temperature of the body for several hours. After repeated experiments of this kind, apparently conducted with great care, Reaumur reached the conclusion that "the gastric juice has no more effect out of the living body in dissolving or digesting the food than water, mucilage, milk, or any other bland fluid."[3] Just why all of these experiments failed to demonstrate a fact so simple does not appear; but to Spallanzani, at least, they were by no means conclusive, and he proceeded to elaborate upon the experiments of Reaumur. He made his experiments in scaled tubes exposed to a certain degree of heat, and showed conclusively that the chemical process does go on, even when the food and gastric juice are removed from their natural environment in the stomach. In this he was opposed by many physiologists, among them John Hunter, but the truth of his demonstrations could not be shaken, and in later years we find Hunter himself completing Spallanzani's experiments by his studies of the post–mortem action of the gastric juice upon the stomach walls.

That Spallanzani's and Hunter's theories of the action of the gastric juice were not at once universally accepted is shown by an essay written by a learned physician in 1834. In speaking of some of Spallanzani's demonstrations, he writes: "In some of the experiments, in order to give the flesh or grains steeped in the gastric juice the same temperature with the body, the phials were introduced under the armpits. But this is not a fair mode of ascertaining the effects of the gastric juice out of the body; for the influence which life may be supposed to have on the solution of the food would be secured in this case. The affinities connected with life would extend to substances in contact with any part of the system: substances placed under the armpits are not placed at least in the same circumstances with those unconnected with a living animal." But just how this writer reaches the conclusion that "the experiments of Reaumur and Spallanzani give no evidence that the gastric juice has any peculiar influence more than water or any other bland fluid in digesting the food"[4] is difficult to understand.

The concluding touches were given to the new theory of digestion by John Hunter, who, as we have seen, at first opposed Spallanzani, but who finally became an ardent champion of the chemical theory. Hunter now carried Spallanzani's experiments further and proved the action of the digestive fluids after death. For many years anatomists had been puzzled by pathological lesion of the stomach, found post mortem, when no symptoms of any disorder of the stomach had been evinced during life. Hunter rightly conceived that these lesions were caused by the action of the gastric juice, which, while unable to act upon the living tissue, continued its action chemically after death, thus digesting the walls of the stomach in which it had been formed. And, as usual with his observations, be turned this discovery to practical use in accounting for certain phenomena of digestion. The following account of the stomach being digested after death was written by Hunter at the desire of Sir John Pringle, when he was president of the Royal Society, and the circumstance which led to this is as follows: "I was opening, in his presence, the body of a patient of his own, where the stomach was in part dissolved, which appeared to him very unaccountable, as there had been no previous symptom that could have led him to suspect any disease in the stomach. I took that opportunity of giving him my ideas respecting it, and told him that I had long been making experiments on digestion, and considered this as one of the facts which proved a converting power in the gastric juice. . . . There are a great many powers in nature which the living principle does not enable the animal matter, with which it is combined, to resist—viz., the mechanical and most of the strongest chemical solvents. It renders it, however, capable of resisting the powers of fermentation, digestion, and perhaps several others, which are well known to act on the same matter when deprived of the living principle and entirely to decompose it. "

Hunter concludes his paper with the following paragraph: "These appearances throw considerable light on the principle of digestion, and show that it is neither a mechanical power, nor contractions of the stomach, nor heat, but something secreted in the coats of the stomach, and thrown into its cavity, which there animalizes the food or assimilates it to the nature of the blood. The power of this juice is confined or limited to certain substances, especially of the vegetable and animal kingdoms; and although this menstruum is capable of acting independently of the stomach, yet it is indebted to that viscus for its continuance.[5]

THE FUNCTION OF RESPIRATION

It is a curious commentary on the crude notions of mechanics of previous generations that it should have been necessary to prove by experiment that the thin, almost membranous stomach of a mammal has not the power to pulverize, by mere attrition, the foods that are taken into it. However, the proof was now for the first time forthcoming, and the question of the general character of the function of digestion was forever set at rest. Almost simultaneously with this great advance, corresponding progress was made in an allied field: the mysteries of respiration were at last cleared up, thanks to the new knowledge of chemistry. The solution of the problem followed almost as a matter of course upon the advances of that science in the latter part of the century. Hitherto no one since Mayow, of the previous century, whose flash of insight had been strangely overlooked and forgotten, had even vaguely surmised the true function of the lungs. The great Boerhaave had supposed that respiration is chiefly important as an aid to the circulation of the blood; his great pupil, Haller, had believed to the day of his death in 1777 that the main purpose of the function is to form the voice. No genius could hope to fathom the mystery of the lungs so long as air was supposed to be a simple element, serving a mere mechanical purpose in the economy of the earth.

But the discovery of oxygen gave the clew, and very soon all the chemists were testing the air that came from the lungs—Dr. Priestley, as usual, being in the van. His initial experiments were made in 1777, and from the outset the problem was as good as solved. Other experimenters confirmed his results in all their essentials—notably Scheele and Lavoisier and Spallanzani and Davy. It was clearly established that there is chemical action in the contact of the air with the tissue of the lungs; that some of the oxygen of the air disappears, and that carbonic–acid gas is added to the inspired air. It was shown, too, that the blood, having come in contact with the air, is changed from black to red in color. These essentials were not in dispute from the first. But as to just what chemical changes caused these results was the subject of controversy. Whether, for example, oxygen is actually absorbed into the blood, or whether it merely unites with carbon given off from the blood, was long in dispute.

Each of the main disputants was biased by his own particular views as to the moot points of chemistry. Lavoisier, for example, believed oxygen gas to be composed of a metal oxygen combined with the alleged element heat; Dr. Priestley thought it a compound of positive electricity and phlogiston; and Humphry Davy, when he entered the lists a little later, supposed it to be a compound of oxygen and light. Such mistaken notions naturally complicated matters and delayed a complete understanding of the chemical processes of respiration. It was some time, too, before the idea gained acceptance that the most important chemical changes do not occur in the lungs themselves, but in the ultimate tissues. Indeed, the matter was not clearly settled at the close of the century. Nevertheless, the problem of respiration had been solved in its essentials. Moreover, the vastly important fact had been established that a process essentially identical with respiration is necessary to the existence not only of all creatures supplied with lungs, but to fishes, insects, and even vegetables—in short, to every kind of living organism.

ERASMUS DARWIN AND VEGETABLE PHYSIOLOGY

Some interesting experiments regarding vegetable respiration were made just at the close of the century by Erasmus Darwin, and recorded in his Botanic Garden as a foot–note to the verse:

"While spread in air the leaves respiring play."

These notes are worth quoting at some length, as they give a clear idea of the physiological doctrines of the time (1799), while taking advance ground as to the specific matter in question:

"There have been various opinions," Darwin says, "concerning the use of the leaves of plants in the vegetable economy. Some have contended that they are perspiratory organs. This does not seem probable from an experiment of Dr. Hales, Vegetable Statics, p. 30. He, found, by cutting off branches of trees with apples on them and taking off the leaves, that an apple exhaled about as much as two leaves the surfaces of which were nearly equal to the apple; whence it would appear that apples have as good a claim to be termed perspiratory organs as leaves. Others have believed them excretory organs of excrementitious juices, but as the vapor exhaled from vegetables has no taste, this idea is no more probable than the other; add to this that in most weathers they do not appear to perspire or exhale at all.

"The internal surface of the lungs or air—vessels in men is said to be equal to the external surface of the whole body, or almost fifteen square feet; on this surface the blood is exposed to the influence of the respired air through the medium, however, of a thin pellicle; by this exposure to the air it has its color changed from deep red to bright scarlet, and acquires something so necessary to the existence of life that we can live scarcely a minute without this wonderful process.

"The analogy between the leaves of plants and the lungs or gills of animals seems to embrace so many circumstances that we can scarcely withhold our consent to their performing similar offices.

"1. The great surface of leaves compared to that of the trunk and branches of trees is such that it would seem to be an organ well adapted for the purpose of exposing the vegetable juices to the influence of the air; this, however, we shall see afterwards is probably performed only by their upper surfaces, yet even in this case the surface of the leaves in general bear a greater proportion to the surface of the tree than the lungs of animals to their external surfaces.

"2. In the lung of animals the blood, after having been exposed to the air in the extremities of the pulmonary artery, is changed in color from deep red to bright scarlet, and certainly in some of its essential properties it is then collected by the pulmonary vein and returned to the heart. To show a similarity of circumstances in the leaves of plants, the following experiment was made, June 24, 1781. A stalk with leaves and seed–vessels of large spurge (Euphorbia helioscopia) had been several days placed in a decoction of madder (Rubia tinctorum) so that the lower part of the stem and two of the undermost leaves were immersed in it. After having washed the immersed leaves in clear water I could readily discover the color of the madder passing along the middle rib of each leaf. The red artery was beautifully visible on the under and on the upper surface of the leaf; but on the upper side many red branches were seen going from it to the extremities of the leaf, which on the other side were not visible except by looking through it against the light. On this under side a system of branching vessels carrying a pale milky fluid were seen coming from the extremities of the leaf, and covering the whole under side of it, and joining two large veins, one on each side of the red artery in the middle rib of the leaf, and along with it descending to the foot–stalk or petiole. On slitting one of these leaves with scissors, and having a magnifying–glass ready, the milky blood was seen oozing out of the returning veins on each side of the red artery in the middle rib, but none of the red fluid from the artery.

"All these appearances were more easily seen in a leaf of Picris treated in the same manner; for in this milky plant the stems and middle rib of the leaves are sometimes naturally colored reddish, and hence the color of the madder seemed to pass farther into the ramifications of their leaf–arteries, and was there beautifully visible with the returning branches of milky veins on each side."

Darwin now goes on to draw an incorrect inference from his observations:

"3. From these experiments," he says, "the upper surface of the leaf appeared to be the immediate organ of respiration, because the colored fluid was carried to the extremities of the leaf by vessels most conspicuous on the upper surface, and there changed into a milky fluid, which is the blood of the plant, and then returned by concomitant veins on the under surface, which were seen to ooze when divided with scissors, and which, in Picris, particularly, render the under surface of the leaves greatly whiter than the upper one."

But in point of fact, as studies of a later generation were to show, it is the under surface of the leaf that is most abundantly provided with stomata, or "breathing–pores." From the stand–point of this later knowledge, it is of interest to follow our author a little farther, to illustrate yet more fully the possibility of combining correct observations with a faulty inference.

"4. As the upper surface of leaves constitutes the organ of respiration, on which the sap is exposed in the termination of arteries beneath a thin pellicle to the action of the atmosphere, these surfaces in many plants strongly repel moisture, as cabbage leaves, whence the particles of rain lying over their surfaces without touching them, as observed by Mr. Melville (Essays Literary and Philosophical: Edinburgh), have the appearance of globules of quicksilver. And hence leaves with the upper surfaces on water wither as soon as in the dry air, but continue green for many days if placed with the under surface on water, as appears in the experiments of Monsieur Bonnet (Usage des Feuilles). Hence some aquatic plants, as the water–lily (Nymphoea), have the lower sides floating on the water, while the upper surfaces remain dry in the air.

"5. As those insects which have many spiracula, or breathing apertures, as wasps and flies, are immediately suffocated by pouring oil upon them, I carefully covered with oil the surfaces of several leaves of phlomis, of Portugal laurel, and balsams, and though it would not regularly adhere, I found them all die in a day or two.

"It must be added that many leaves are furnished with muscles about their foot–stalks, to turn their surfaces to the air or light, as mimosa or Hedysarum gyrans. From all these analogies I think there can be no doubt but that leaves of trees are their lungs, giving out a phlogistic material to the atmosphere, and absorbing oxygen, or vital air.

"6. The great use of light to vegetation would appear from this theory to be by disengaging vital air from the water which they perspire, and thence to facilitate its union with their blood exposed beneath the thin surface of their leaves; since when pure air is thus applied it is probable that it can be more readily absorbed. Hence, in the curious experiments of Dr. Priestley and Mr. Ingenhouz, some plants purified less air than others—that is, they perspired less in the sunshine; and Mr. Scheele found that by putting peas into water which about half covered them they converted the vital air into fixed air, or carbonic–acid gas, in the same manner as in animal respiration.

"7. The circulation in the lungs or leaves of plants is very similar to that of fish. In fish the blood, after having passed through their gills, does not return to the heart as from the lungs of air–breathing animals, but the pulmonary vein taking the structure of an artery after having received the blood from the gills, which there gains a more florid color, distributes it to the other parts of their bodies. The same structure occurs in the livers of fish, whence we see in those animals two circulations independent of the power of the heart—viz., that beginning at the termination of the veins of the gills and branching through the muscles, and that which passes through the liver; both which are carried on by the action of those respective arteries and veins."[6]

Darwin is here a trifle fanciful in forcing the analogy between plants and animals. The circulatory system of plants is really not quite so elaborately comparable to that of fishes as he supposed. But the all−important idea of the uniformity underlying the seeming diversity of Nature is here exemplified, as elsewhere in the writings of Erasmus Darwin; and, more specifically, a clear grasp of the essentials of the function of respiration is fully demonstrated.

ZOOLOGY AT THE CLOSE OF THE EIGHTEENTH CENTURY

Several causes conspired to make exploration all the fashion during the closing epoch of the eighteenth century. New aid to the navigator had been furnished by the perfected compass and quadrant, and by the invention of the chronometer; medical science had banished scurvy, which hitherto had been a perpetual menace to the voyager; and, above all, the restless spirit of the age impelled the venturesome to seek novelty in fields altogether new. Some started for the pole, others tried for a northeast or northwest passage to India, yet others sought the great fictitious antarctic continent told of by tradition. All these of course failed of their immediate purpose, but they added much to the world's store of knowledge and its fund of travellers' tales.

Among all these tales none was more remarkable than those which told of strange living creatures found in antipodal lands. And here, as did not happen in every field, the narratives were often substantiated by the exhibition of specimens that admitted no question. Many a company of explorers returned more or less laden with such trophies from the animal and vegetable kingdoms, to the mingled astonishment, delight, and bewilderment of the closet naturalists. The followers of Linnaeus in the "golden age of natural history," a few decades before, had increased the number of known species of fishes to about four hundred, of birds to one thousand, of insects to three thousand, and of plants to ten thousand. But now these sudden accessions from new territories doubled the figure for plants, tripled it for fish and birds, and brought the number of described insects above twenty thousand. Naturally enough, this wealth of new material was sorely puzzling to the classifiers. The more discerning began to see that the artificial system of Linnaeus, wonderful and useful as it had been, must be advanced upon before the new material could be satisfactorily disposed of. The way to a more natural system, based on less arbitrary signs, had been pointed out by Jussieu in botany, but the zoologists were not prepared to make headway towards such a system until they should gain a wider understanding of the organisms with which they had to deal through comprehensive studies of anatomy. Such studies of individual forms in their relations to the entire scale of organic beings were pursued in these last decades of the century, but though two or three most important generalizations were achieved (notably Kaspar Wolff's conception of the cell as the basis of organic life, and Goethe's all—important doctrine of metamorphosis of parts), yet, as a whole, the work of the anatomists of the period was germinative rather than fruit—bearing. Bichat's volumes, telling of the recognition of the fundamental tissues of the body, did not begin to appear till the last year of the century. The announcement by Cuvier of the doctrine of correlation of parts bears the same date, but in general the studies of this great naturalist, which in due time were to stamp him as the successor of Linnaeus, were as yet only fairly begun.

V. ANATOMY AND PHYSIOLOGY IN THE NINETEENTH CENTURY

CUVIER AND THE CORRELATION OF PARTS

We have seen that the focal points of the physiological world towards the close of the eighteenth century were Italy and England, but when Spallanzani and Hunter passed away the scene shifted to France. The time was peculiarly propitious, as the recent advances in many lines of science had brought fresh data for the student of animal life which were in need of classification, and, as several minds capable of such a task were in the field, it was natural that great generalizations should have come to be quite the fashion. Thus it was that Cuvier came forward with a brand—new classification of the animal kingdom, establishing four great types of being, which he called vertebrates, mollusks, articulates, and radiates. Lamarck had shortly before established the broad distinction between animals with and those without a backbone; Cuvier's Classification divided the latter—the invertebrates—into three minor groups. And this division, familiar ever since to all students of zoology, has only in very recent years been supplanted, and then not by revolution, but by a further division, which the elaborate recent studies of lower forms of life seemed to make desirable.

In the course of those studies of comparative anatomy which led to his new classification, Cuvier's attention was called constantly to the peculiar co-ordination of parts in each individual organism. Thus an animal with sharp talons for catching living prey—as a member of the cat tribe—has also sharp teeth, adapted for tearing up the flesh of its victim, and a particular type of stomach, quite different from that of herbivorous creatures. This adaptation of all the parts of the animal to one another extends to the most diverse parts of the organism, and enables the skilled anatomist, from the observation of a single typical part, to draw inferences as to the structure of the entire animal—a fact which was of vast aid to Cuvier in his studies of paleontology. It did not enable Cuvier, nor does it enable any one else, to reconstruct fully the extinct animal from observation of a single bone, as has sometimes been asserted, but what it really does establish, in the hands of an expert, is sufficiently astonishing.

"While the study of the fossil remains of the greater quadrupeds is more satisfactory," he writes, "by the clear results which it affords, than that of the remains of other animals found in a fossil state, it is also complicated with greater and more numerous difficulties. Fossil shells are usually found quite entire, and retaining all the characters requisite for comparing them with the specimens contained in collections of natural history, or represented in the works of naturalists. Even the skeletons of fishes are found more or less entire, so that the general forms of their bodies can, for the most part, be ascertained, and usually, at least, their generic and specific characters are determinable, as these are mostly drawn from their solid parts. In quadrupeds, on the contrary, even when their entire skeletons are found, there is great difficulty in discovering their distinguishing characters, as these are chiefly founded upon their hairs and colors and other marks which have disappeared previous to their incrustation. It is also very rare to find any fossil skeletons of quadrupeds in any degree approaching to a complete state, as the strata for the most part only contain separate bones, scattered confusedly and almost always broken and reduced to fragments, which are the only means left to naturalists for ascertaining the species or genera to which they have belonged.

"Fortunately comparative anatomy, when thoroughly understood, enables us to surmount all these difficulties, as a careful application of its principles instructs us in the correspondences and dissimilarities of the forms of organized bodies of different kinds, by which each may be rigorously ascertained from almost every fragment of its various parts and organs.

"Every organized individual forms an entire system of its own, all the parts of which naturally correspond, and concur to produce a certain definite purpose, by reciprocal reaction, or by combining towards the same end. Hence none of these separate parts can change their forms without a corresponding change in the other parts of the same animal, and consequently each of these parts, taken separately, indicates all the other parts to which it has belonged. Thus, as I have elsewhere shown, if the viscera of an animal are so organized as only to be fitted for the digestion of recent flesh, it is also requisite that the jaws should be so constructed as to fit them for devouring prey; the claws must be constructed for seizing and tearing it to pieces; the teeth for cutting and dividing its flesh; the entire system of the limbs, or organs of motion, for pursuing and overtaking it; and the organs of sense for discovering it at a distance. Nature must also have endowed the brain of the animal with instincts sufficient for concealing itself and for laying plans to catch its necessary victims. . . . .

"To enable the animal to carry off its prey when seized, a corresponding force is requisite in the muscles which elevate the head, and this necessarily gives rise to a determinate form of the vertebrae to which these muscles are attached and of the occiput into which they are inserted. In order that the teeth of a carnivorous animal may be able to cut the flesh, they require to be sharp, more or less so in proportion to the greater or less quantity of flesh that they have to cut. It is requisite that their roots should be solid and strong, in proportion to the quantity and size of the bones which they have to break to pieces. The whole of these circumstances must necessarily influence the development and form of all the parts which contribute to move the jaws. . . . .

After these observations, it will be easily seen that similar conclusions may be drawn with respect to the limbs of carnivorous animals, which require particular conformations to fit them for rapidity of motion in general; and that similar considerations must influence the forms and connections of the vertebrae and other bones constituting the trunk of the body, to fit them for flexibility and readiness of motion in all directions. The bones also of the nose, of the orbit, and of the ears require certain forms and structures to fit them for giving perfection to the senses of smell, sight, and hearing, so necessary to animals of prey. In short, the shape and structure of the teeth regulate the forms of the condyle, of the shoulder—blade, and of the claws, in the same manner as the equation of a curve regulates all its other properties; and as in regard to any particular curve all its properties may be ascertained by assuming each separate property as the foundation of a particular equation, in the same manner a claw, a shoulder–blade, a condyle, a leg or arm bone, or any other bone separately considered, enables us to discover the description of teeth to which they have belonged; and so also reciprocally we may determine the forms of the other bones from the teeth. Thus commencing our investigations by a careful survey of any one bone by itself, a person who is sufficiently master of the laws of organic structure may, as it were, reconstruct the whole animal to which that bone belonged."[1]

We have already pointed out that no one is quite able to perform the necromantic feat suggested in the last sentence; but the exaggeration is pardonable in the enthusiast to whom the principle meant so much and in whose hands it extended so far.

Of course this entire principle, in its broad outlines, is something with which every student of anatomy had been familiar from the time when anatomy was first studied, but the full expression of the "law of co–ordination," as Cuvier called it, had never been explicitly made before; and, notwithstanding its seeming obviousness, the exposition which Cuvier made of it in the introduction to his classical work on comparative anatomy, which was published during the first decade of the nineteenth century, ranks as a great discovery. It is one of those generalizations which serve as guideposts to other discoveries.

BICHAT AND THE BODILY TISSUES

Much the same thing may be said of another generalization regarding the animal body, which the brilliant young French physician Marie Francois Bichat made in calling attention to the fact that each vertebrate organism, including man, has really two quite different sets of organs—one set under volitional control, and serving the end of locomotion, the other removed from volitional control, and serving the ends of the "vital processes" of digestion, assimilation, and the like. He called these sets of organs the animal system and the organic system, respectively. The division thus pointed out was not quite new, for Grimaud, professor of physiology in the University of Montpellier, had earlier made what was substantially the same classification of the functions into "internal or digestive and external or locomotive"; but it was Bichat's exposition that gave currency to the idea.

Far more important, however, was another classification which Bichat put forward in his work on anatomy, published just at the beginning of the last century. This was the division of all animal structures into what Bichat called tissues, and the pointing out that there are really only a few kinds of these in the body, making up all the diverse organs. Thus muscular organs form one system; membranous organs another; glandular organs a third; the vascular mechanism a fourth, and so on. The distinction is so obvious that it seems rather difficult to conceive that it could have been overlooked by the earliest anatomists; but, in point of fact, it is only obvious because now it has been familiarly taught for almost a century. It had never been given explicit expression before the time of Bichat, though it is said that Bichat himself was somewhat indebted for it to his master, Desault, and to the famous alienist Pinel.

However that may be, it is certain that all subsequent anatomists have found Bichat's classification of the tissues of the utmost value in their studies of the animal functions. Subsequent advances were to show that the distinction between the various tissues is not really so fundamental as Bichat supposed, but that takes nothing from the practical value of the famous classification.

It was but a step from this scientific classification of tissues to a similar classification of the diseases affecting them, and this was one of the greatest steps towards placing medicine on the plane of an exact science. This subject of these branches completely fascinated Bichat, and he exclaimed, enthusiastically: "Take away some fevers and nervous trouble, and all else belongs to the kingdom of pathological anatomy." But out of this enthusiasm came great results. Bichat practised as he preached, and, believing that it was only possible to understand disease by observing the symptoms carefully at the bedside, and, if the disease terminated fatally, by post—mortem examination, he was so arduous in his pursuit of knowledge that within a period of less than six months he had made over six hundred autopsies—a record that has seldom, if ever, been equalled. Nor were his efforts fruitless, as a single example will suffice to show. By his examinations he was able to prove that diseases of the chest, which had formerly been classed under the indefinite name "peripneumonia," might involve three different structures, the pleural sac covering the lungs, the lung itself, and the bronchial tubes, the diseases affecting these organs being known respectively as pleuritis, pneumonia, and bronchitis, each one differing from the others as to prognosis and treatment. The advantage of such an exact classification needs no demonstration.

LISTER AND THE PERFECTED MICROSCOPE

At the same time when these broad macroscopical distinctions were being drawn there were other workers who were striving to go even deeper into the intricacies of the animal mechanism with the aid of the microscope. This undertaking, however, was beset with very great optical difficulties, and for a long time little advance was made upon the work of preceding generations. Two great optical barriers, known technically as spherical and chromatic aberration—the one due to a failure of the rays of light to fall all in one plane when focalized through a lens, the other due to the dispersive action of the lens in breaking the white light into prismatic colors—confronted the makers of microscopic lenses, and seemed all but insuperable. The making of achromatic lenses for telescopes had been accomplished, it is true, by Dolland in the previous century, by the union of lenses of crown glass with those of flint glass, these two materials having different indices of refraction and dispersion. But, aside from the mechanical difficulties which arise when the lens is of the minute dimensions required for use with the microscope, other perplexities are introduced by the fact that the use of a wide pencil of light is a desideratum, in order to gain sufficient illumination when large magnification is to be secured.

In the attempt to overcome those difficulties, the foremost physical philosophers of the time came to the aid of the best opticians. Very early in the century, Dr. (afterwards Sir David) Brewster, the renowned Scotch physicist, suggested that certain advantages might accrue from the use of such gems as have high refractive and low dispersive indices, in place of lenses made of glass. Accordingly lenses were made of diamond, of sapphire, and so on, and with some measure of success. But in 1812 a much more important innovation was introduced by Dr. William Hyde Wollaston, one of the greatest and most versatile, and, since the death of Cavendish, by far the most eccentric of English natural philosophers. This was the suggestion to use two plano–convex lenses, placed at a prescribed distance apart, in lieu of the single double–convex lens generally used. This combination largely overcame the spherical aberration, and it gained immediate fame as the "Wollaston doublet."

To obviate loss of light in such a doublet from increase of reflecting surfaces, Dr. Brewster suggested filling the interspace between the two lenses with a cement having the same index of refraction as the lenses themselves—an improvement of manifest advantage. An improvement yet more important was made by Dr. Wollaston himself in

66

the introduction of the diaphragm to limit the field of vision between the lenses, instead of in front of the anterior lens. A pair of lenses thus equipped Dr. Wollaston called the periscopic microscope. Dr. Brewster suggested that in such a lens the same object might be attained with greater ease by grinding an equatorial groove about a thick or globular lens and filling the groove with an opaque cement. This arrangement found much favor, and came subsequently to be known as a Coddington lens, though Mr. Coddington laid no claim to being its inventor.

Sir John Herschel, another of the very great physicists of the time, also gave attention to the problem of improving the microscope, and in 1821 he introduced what was called an aplanatic combination of lenses, in which, as the name implies, the spherical aberration was largely done away with. It was thought that the use of this Herschel aplanatic combination as an eyepiece, combined with the Wollaston doublet for the objective, came as near perfection as the compound microscope was likely soon to come. But in reality the instrument thus constructed, though doubtless superior to any predecessor, was so defective that for practical purposes the simple microscope, such as the doublet or the Coddington, was preferable to the more complicated one.

Many opticians, indeed, quite despaired of ever being able to make a satisfactory refracting compound microscope, and some of them had taken up anew Sir Isaac Newton's suggestion in reference to a reflecting microscope. In particular, Professor Giovanni Battista Amici, a very famous mathematician and practical optician of Modena, succeeded in constructing a reflecting microscope which was said to be superior to any compound microscope of the time, though the events of the ensuing years were destined to rob it of all but historical value. For there were others, fortunately, who did not despair of the possibilities of the refracting microscope, and their efforts were destined before long to be crowned with a degree of success not even dreamed of by any preceding generation.

The man to whom chief credit is due for directing those final steps that made the compound microscope a practical implement instead of a scientific toy was the English amateur optician Joseph Jackson Lister. Combining mathematical knowledge with mechanical ingenuity, and having the practical aid of the celebrated optician Tulley, he devised formulae for the combination of lenses of crown glass with others of flint glass, so adjusted that the refractive errors of one were corrected or compensated by the other, with the result of producing lenses of hitherto unequalled powers of definition; lenses capable of showing an image highly magnified, yet relatively free from those distortions and fringes of color that had heretofore been so disastrous to true interpretation of magnified structures.

Lister had begun his studies of the lens in 1824, but it was not until 1830 that he contributed to the Royal Society the famous paper detailing his theories and experiments. Soon after this various continental opticians who had long been working along similar lines took the matter up, and their expositions, in particular that of Amici, introduced the improved compound microscope to the attention of microscopists everywhere. And it required but the most casual trial to convince the experienced observers that a new implement of scientific research had been placed in their hands which carried them a long step nearer the observation of the intimate physical processes which lie at the foundation of vital phenomena. For the physiologist this perfection of the compound microscope had the same significance that the, discovery of America had for the fifteenth-century geographers—it promised a veritable world of utterly novel revelations. Nor was the fulfilment of that promise long delayed.

Indeed, so numerous and so important were the discoveries now made in the realm of minute anatomy that the rise of histology to the rank of an independent science may be said to date from this period. Hitherto, ever since the discovery of magnifying-glasses, there had been here and there a man, such as Leuwenhoek or Malpighi, gifted with exceptional vision, and perhaps unusually happy in his conjectures, who made important contributions to the knowledge of the minute structure of organic tissues; but now of a sudden it became possible for the veriest tyro to confirm or refute the laborious observations of these pioneers, while the skilled observer could step easily beyond the barriers of vision that hitherto were quite impassable. And so, naturally enough, the physiologists of the fourth decade of the nineteenth century rushed as eagerly into the new realm of the microscope as, for example, their successors of to-day are exploring the realm of the X-ray.

Lister himself, who had become an eager interrogator of the instrument he had perfected, made many important discoveries, the most notable being his final settlement of the long-mooted question as to the true form of the red corpuscles of the human blood. In reality, as everybody knows nowadays, these are biconcave disks, but owing to their peculiar figure it is easily possible to misinterpret the appearances they present when seen through a poor lens, and though Dr. Thomas Young and various other observers had come very near the truth regarding them, unanimity of opinion was possible only after the verdict of the perfected microscope was given.

These blood corpuscles are so infinitesimal in size that something like five millions of them are found in each cubic millimetre of the blood, yet they are isolated particles, each having, so to speak, its own personality. This, of course, had been known to microscopists since the days of the earliest lenses. It had been noticed, too, by here and there an observer, that certain of the solid tissues seemed to present something of a granular texture, as if they, too, in their ultimate constitution, were made up of particles. And now, as better and better lenses were constructed, this idea gained ground constantly, though for a time no one saw its full significance. In the case of vegetable tissues, indeed, the fact that little particles encased a membranous covering, and called cells, are the ultimate visible units of structure had long been known. But it was supposed that animal tissues differed radically from this construction. The elementary particles of vegetables "were regarded to a certain extent as individuals which composed the entire plant, while, on the other hand, no such view was taken of the elementary parts of animals."

ROBERT BROWN AND THE CELL NUCLEUS

In the year 1833 a further insight into the nature of the ultimate particles of plants was gained through the observation of the English microscopist Robert Brown, who, in the course of his microscopic studies of the epidermis of orchids, discovered in the cells "an opaque spot," which he named the nucleus. Doubtless the same "spot" had been seen often enough before by other observers, but Brown was the first to recognize it as a component part of the vegetable cell and to give it a name.

"I shall conclude my observations on Orchideae," said Brown, "with a notice of some points of their general structure, which chiefly relate to the cellular tissue. In each cell of the epidermis of a great part of this family, especially of those with membranous leaves, a single circular areola, generally somewhat more opaque than, the membrane of the cell, is observable. This areola, which is more or less distinctly granular, is slightly convex, and although it seems to be on the surface is in reality covered by the outer lamina of the cell. There is no regularity as to its place in the cell; it is not unfrequently, however, central or nearly so.

"As only one areola belongs to each cell, and as in many cases where it exists in the common cells of the epidermis, it is also visible in the cutaneous glands or stomata, and in these is always double—one being on each side of the limb—it is highly probable that the cutaneous gland is in all cases composed of two cells of peculiar form, the line of union being the longitudinal axis of the disk or pore.

"This areola, or nucleus of the cell as perhaps it might be termed, is not confined to the epidermis, being also found, not only in the pubescence of the surface, particularly when jointed, as in cypripedium, but in many cases in the parenchyma or internal cells of the tissue, especially when these are free from the deposition of granular matter.

"In the compressed cells of the epidermis the nucleus is in a corresponding degree flattened; but in the internal tissue it is often nearly spherical, more or less firmly adhering to one of the walls, and projecting into the cavity of the cell. In this state it may not unfrequently be found. in the substance of the column and in that of the perianthium.

"The nucleus is manifest also in the tissue of the stigma, where in accordance with the compression of the utriculi, it has an intermediate form, being neither so much flattened as in the epidermis nor so convex as it is in the internal tissue of the column.

"I may here remark that I am acquainted with one case of apparent exception to the nucleus being solitary in each utriculus or cell—namely, in Bletia Tankervilliae. In the utriculi of the stigma of this plant, I have generally, though not always, found a second areola apparently on the surface, and composed of much larger granules than the ordinary nucleus, which is formed of very minute granular matter, and seems to be deep seated.

"Mr. Bauer has represented the tissue of the stigma, in the species of Bletia, both before and, as he believes, after impregnation; and in the latter state the utriculi are marked with from one to three areolae of similar appearance.

"The nucleus may even be supposed to exist in the pollen of this family. In the early stages of its formation, at least a minute areola is of ten visible in the simple grain, and in each of the constituent parts of cells of the compound grain. But these areolae may perhaps rather be considered as merely the points of production of the tubes.

"This nucleus of the cell is not confined to orchideae, but is equally manifest in many other monocotyledonous families; and I have even found it, hitherto however in very few cases, in the epidermis of dicotyledonous plants; though in this primary division it may perhaps be said to exist in the early stages of development of the pollen. Among monocotyledons, the orders in which it is most remarkable are Liliaceae, Hemerocallideae, Asphodeleae, Irideae, and Commelineae.

"In some plants belonging to this last–mentioned family, especially in Tradascantia virginica, and several nearly related species, it is uncommonly distinct, not in the epidermis and in the jointed hairs of the filaments, but in the tissue of the stigma, in the cells of the ovulum even before impregnation, and in all the stages of formation of the grains of pollen, the evolution of which is so remarkable in tradascantia.

"The few indications of the presence of this nucleus, or areola, that I have hitherto met with in the publications of botanists are chiefly in some figures of epidermis, in the recent works of Meyen and Purkinje, and in one case, in M. Adolphe Broigniart's memoir on the structure of leaves. But so little importance seems to be attached to it that the appearance is not always referred to in the explanations of the figures in which it is represented. Mr. Bauer, however, who has also figured it in the utriculi of the stigma of Bletia Tankervilliae has more particularly noticed it, and seems to consider it as only visible after impregnation."[2]

SCHLEIDEN AND SCHWANN AND THE CELL THEORY

That this newly recognized structure must be important in the economy of the cell was recognized by Brown himself, and by the celebrated German Meyen, who dealt with it in his work on vegetable physiology, published not long afterwards; but it remained for another German, the professor of botany in the University of Jena, Dr. M. J. Schleiden, to bring the nucleus to popular attention, and to assert its all–importance in the economy of the cell.

Schleiden freely acknowledged his indebtedness to Brown for first knowledge of the nucleus, but he soon carried his studies of that structure far beyond those of its discoverer. He came to believe that the nucleus is really the most important portion of the cell, in that it is the original structure from which the remainder of the cell is developed. Hence he named it the cytoblast. He outlined his views in an epochal paper published in

Muller's Archives in 1838, under title of "Beitrage zur Phytogenesis." This paper is in itself of value, yet the most important outgrowth of Schleiden's observations of the nucleus did not spring from his own labors, but from those of a friend to whom he mentioned his discoveries the year previous to their publication. This friend was Dr. Theodor Schwann, professor of physiology in the University of Louvain.

At the moment when these observations were communicated to him Schwann was puzzling over certain details of animal histology which he could not clearly explain. His great teacher, Johannes Muller, had called attention to the strange resemblance to vegetable cells shown by certain cells of the chorda dorsalis (the embryonic cord from which the spinal column is developed), and Schwann himself had discovered a corresponding similarity in the branchial cartilage of a tadpole. Then, too, the researches of Friedrich Henle had shown that the particles that make up the epidermis of animals are very cell–like in appearance. Indeed, the cell–like character of certain animal tissues had come to be matter of common note among students of minute anatomy. Schwann felt that this similarity could not be mere coincidence, but he had gained no clew to further insight until Schleiden called his attention to the nucleus. Then at once he reasoned that if there really is the correspondence between vegetable and animal tissues that he suspected, and if the nucleus is so important in the vegetable cell as Schleiden believed, the nucleus should also be found in the ultimate particles of animal tissues.

Schwann's researches soon showed the entire correctness of this assumption. A closer study of animal tissues under the microscope showed, particularly in the case of embryonic tissues, that "opaque spots" such as Schleiden described are really to be found there in abundance—forming, indeed, a most characteristic phase of the structure. The location of these nuclei at comparatively regular intervals suggested that they are found in definite compartments of the tissue, as Schleiden had shown to be the case with vegetables; indeed, the walls that separated such cell–like compartments one from another were in some cases visible. Particularly was this found to be the case with embryonic tissues, and the study of these soon convinced Schwann that his original surmise had been correct, and that all animal tissues are in their incipiency composed of particles not unlike the ultimate particles of vegetables in short, of what the botanists termed cells. Adopting this name, Schwann propounded what soon became famous as his cell theory, under title of Mikroskopische Untersuchungen uber die Ubereinstimmung in der Structur und dent Wachsthum der Thiere und Pflanzen. So expeditious had been his work that this book was published early in 1839, only a few months after the appearance of Schleiden's paper.

As the title suggests, the main idea that actuated Schwann was to unify vegetable and animal tissues. Accepting cell–structure as the basis of all vegetable tissues, he sought to show that the same is true of animal tissues, all the seeming diversities of fibre being but the alteration and development of what were originally simple cells. And by cell Schwann meant, as did Schleiden also, what the word ordinarily implies—a cavity walled in on all sides. He conceived that the ultimate constituents of all tissues were really such minute cavities, the most important part of which was the cell wall, with its associated nucleus. He knew, indeed, that the cell might be filled with fluid contents, but he regarded these as relatively subordinate in importance to the wall itself. This, however, did not apply to the nucleus, which was supposed to lie against the cell wall and in the beginning to generate it. Subsequently the wall might grow so rapidly as to dissociate itself from its contents, thus becoming a hollow bubble or true cell; but the nucleus, as long as it lasted, was supposed to continue in contact with the cell wall. Schleiden had even supposed the nucleus to be a constituent part of the wall, sometimes lying enclosed between two layers of its substance, and Schwann quoted this view with seeming approval. Schwann believed, however, that in the mature cell the nucleus ceased to be functional and disappeared.

The main thesis as to the similarity of development of vegetable and animal tissues and the cellular nature of the ultimate constitution of both was supported by a mass of carefully gathered evidence which a multitude of microscopists at once confirmed, so Schwann's work became a classic almost from the moment of its publication. Of course various other workers at once disputed Schwann's claim to priority of discovery, in particular the English microscopist Valentin, who asserted, not without some show of justice, that he was working closely along the same lines. Put so, for that matter, were numerous others, as Henle, Turpin, Du–mortier, Purkinje, and Muller, all of whom Schwann himself had quoted. Moreover, there were various physiologists who earlier than any of these had foreshadowed the cell theory—notably Kaspar Friedrich Wolff, towards the close of the previous century, and Treviranus about 1807, But, as we have seen in so many other departments of science, it is one thing to foreshadow a discovery, it is quite another to give it full expression and make it germinal of other discoveries. And when Schwann put forward the explicit claim that "there is one universal principle of development for the elementary parts, of organisms, however different, and this principle is the formation of cells," he enunciated a doctrine which was for all practical purposes absolutely new and opened up a novel field for the microscopist to enter. A most important era in physiology dates from the publication of his book in 1839.

THE CELL THEORY ELABORATED

That Schwann should have gone to embryonic tissues for the establishment of his ideas was no doubt due very largely to the influence of the great Russian Karl Ernst von Baer, who about ten years earlier had published the first part of his celebrated work on embryology, and whose ideas were rapidly gaining ground, thanks largely to the advocacy of a few men, notably Johannes Muller, in Germany, and William B. Carpenter, in England, and to the fact that the improved microscope had made minute anatomy popular. Schwann's researches made it plain that the best field for the study of the animal cell is here, and a host of explorers entered the field. The result of their observations was, in the main, to confirm the claims of Schwann as to the universal prevalence of the cell. The long−current idea that animal tissues grow only as a sort of deposit from the blood−vessels was now discarded, and the fact of so−called plantlike growth of animal cells, for which Schwann contended, was universally accepted. Yet the full measure of the affinity between the two classes of cells was not for some time generally apprehended.

Indeed, since the substance that composes the cell walls of plants is manifestly very different from the limiting membrane of the animal cell, it was natural, so long as the, wall was considered the most essential part of the structure, that the divergence between the two classes of cells should seem very pronounced. And for a time this was the conception of the matter that was uniformly accepted. But as time went on many observers had their attention called to the peculiar characteristics of the contents of the cell, and were led to ask themselves whether these might not be more important than had been supposed. In particular, Dr. Hugo von Mohl, professor of botany in the University of Tubingen, in the course of his exhaustive studies of the vegetable cell, was impressed with the peculiar and characteristic appearance of the cell contents. He observed universally within the cell "an opaque, viscid fluid, having granules intermingled in it," which made up the main substance of the cell, and which particularly impressed him because under certain conditions it could be seen to be actively in motion, its parts separated into filamentous streams.

Von Mohl called attention to the fact that this motion of the cell contents had been observed as long ago as 1774 by Bonaventura Corti, and rediscovered in 1807 by Treviranus, and that these observers had described the phenomenon under the "most unsuitable name of 'rotation of the cell sap.' Von Mohl recognized that the streaming substance was something quite different from sap. He asserted that the nucleus of the cell lies within this substance and not attached to the cell wall as Schleiden had contended. He saw, too, that the chlorophyl granules, and all other of the cell contents, are incorporated with the "opaque, viscid fluid," and in 1846 he had become so impressed with the importance of this universal cell substance that be gave it the name of protoplasm. Yet in so doing he had no intention of subordinating the cell wall. The fact that Payen, in 1844, had demonstrated that the cell walls of all vegetables, high or low, are composed largely of one substance, cellulose, tended to strengthen the position of the cell wall as the really essential structure, of which the protoplasmic contents were only subsidiary products.

Meantime, however, the students of animal histology were more and more impressed with the seeming preponderance of cell contents over cell walls in the tissues they studied. They, too, found the cell to be filled with a viscid, slimy fluid capable of motion. To this Dujardin gave the name of sarcode. Presently it came to be known, through the labors of Kolliker, Nageli, Bischoff, and various others, that there are numerous lower forms of animal life which seem to be composed of this sarcode, without any cell wall whatever. The same thing seemed to be true of certain cells of higher organisms, as the blood corpuscles. Particularly in the case of cells that change their shape markedly, moving about in consequence of the streaming of their sarcode, did it seem certain that no cell wall is present, or that, if present, its role must be insignificant.

And so histologists came to question whether, after all, the cell contents rather than the enclosing wall must not be the really essential structure, and the weight of increasing observations finally left no escape from the conclusion that such is really the case. But attention being thus focalized on the cell contents, it was at once apparent that there is a far closer similarity between the ultimate particles of vegetables and those of animals than had been supposed. Cellulose and animal membrane being now regarded as more by-products, the way was clear for the recognition of the fact that vegetable protoplasm and animal sarcode are marvellously similar in appearance and general properties. The closer the observation the more striking seemed this similarity; and finally, about 1860, it was demonstrated by Heinrich de Bary and by Max Schultze that the two are to all intents and purposes identical. Even earlier Remak had reached a similar conclusion, and applied

Von Mohl's word protoplasm to animal cell contents, and now this application soon became universal. Thenceforth this protoplasm was to assume the utmost importance in the physiological world, being recognized as the universal "physical basis of life," vegetable and animal alike. This amounted to the logical extension and culmination of Schwann's doctrine as to the similarity of development of the two animate kingdoms. Yet at the, same time it was in effect the banishment of the cell that Schwann had defined. The word cell was retained, it is true, but it no longer signified a minute cavity. It now implied, as Schultze defined it, "a small mass of protoplasm endowed with the attributes of life." This definition was destined presently to meet with yet another modification, as we shall see; but the conception of the protoplasmic mass as the essential ultimate structure, which might or might not surround itself with a protective covering, was a permanent addition to physiological knowledge. The earlier idea had, in effect, declared the shell the most important part of the egg; this developed view assigned to the yolk its true position.

In one other important regard the theory of Schleiden and Schwann now became modified. This referred to the origin of the cell. Schwann had regarded cell growth as a kind of crystallization, beginning with the deposit of a nucleus about a granule in the intercellular substance—the cytoblastema, as Schleiden called it. But Von Mohl, as early as 1835, had called attention to the formation of new vegetable cells through the division of a pre–existing cell. Ehrenberg, another high authority of the time, contended that no such division occurs, and the matter was still in dispute when Schleiden came forward with his discovery of so–called free cell–formation within the parent cell, and this for a long time diverted attention from the process of division which Von Mohl had described. All manner of schemes of cell–formation were put forward during the ensuing years by a multitude of observers, and gained currency notwithstanding Von Mohl's reiterated contention that there are really but two ways in which the formation of new cells takes place—namely, "first, through division of older cells; secondly, through the formation of secondary cells lying free in the cavity of a cell."

But gradually the researches of such accurate observers as Unger, Nageli, Kolliker, Reichart, and Remak tended to confirm the opinion of Von Mohl that cells spring only from cells, and finally Rudolf Virchow brought the matter to demonstration about 1860. His Omnis cellula e cellula became from that time one of the accepted data of physiology. This was supplemented a little later by Fleming's Omnis nucleus e nucleo, when still more refined methods of observation had shown that the part of the cell which always first undergoes change preparatory to new cell—formation is the all—essential nucleus. Thus the nucleus was restored to the important position which Schwann and Schleiden had given it, but with greatly altered significance. Instead of being a structure generated de novo from non—cellular substance, and disappearing as soon as its function of cell—formation was accomplished, the nucleus was now known as the central and permanent feature of every cell, indestructible while the cell lives, itself the division—product of a pre—existing nucleus, and the parent, by division of its substance, of other generations of nuclei. The word cell received a final definition as "a small mass of protoplasm supplied with a nucleus."

In this widened and culminating general view of the cell theory it became clear that every animate organism, animal or vegetable, is but a cluster of nucleated cells, all of which, in each individual case, are the direct descendants of a single primordial cell of the ovum. In the developed individuals of higher organisms the successive generations of cells become marvellously diversified in form and in specific functions; there is a wonderful division of labor, special functions being chiefly relegated to definite groups of cells; but from first to last there is no function developed that is not present, in a primitive way, in every cell, however isolated; nor does the developed cell, however specialized, ever forget altogether any one of its primordial functions or capacities. All physiology, then, properly interpreted, becomes merely a study of cellular activities; and the development of the cell theory takes its place as the great central generalization in physiology of the nineteenth century. Something of the later developments of this theory we shall see in another connection.

ANIMAL CHEMISTRY

Just at the time when the microscope was opening up the paths that were to lead to the wonderful cell theory, another novel line of interrogation of the living organism was being put forward by a different set of observers. Two great schools of physiological chemistry had arisen—one under guidance of Liebig and Wohler, in Germany, the other dominated by the great French master Jean Baptiste Dumas. Liebig had at one time contemplated the study of medicine, and Dumas had achieved distinction in connection with Prevost, at Geneva, in the field of pure physiology before he turned his attention especially to chemistry. Both these masters, therefore, and Wohler as well, found absorbing interest in those phases of chemistry that have to do with the functions of living tissues; and it was largely through their efforts and the labors of their followers that the prevalent idea that vital processes are dominated by unique laws was discarded and physiology was brought within the recognized province of the chemist. So at about the time when the microscope had taught that the cell is the really essential structure of the living organism, the chemists had come to understand that every function of the organism is really the expression of a chemical change—that each cell is, in short, a miniature chemical laboratory. And it was this combined point of view of anatomist and chemist, this union of hitherto dissociated forces, that made possible the inroads into the unexplored fields of physiology that were effected towards the middle of the nineteenth century.

One of the first subjects reinvestigated and brought to proximal solution was the long—mooted question of the digestion of foods. Spallanzani and Hunter had shown in the previous century that digestion is in some sort a solution of foods; but little advance was made upon their work until 1824, when Prout detected the presence of hydrochloric acid in the gastric juice. A decade later Sprott and Boyd detected the existence of peculiar glands in the gastric mucous membrane; and Cagniard la Tour and Schwann independently discovered that the really active principle of the gastric juice is a substance which was named pepsin, and which was shown by Schwann to be active in the presence of hydrochloric acid.

Almost coincidently, in 1836, it was discovered by Purkinje and Pappenheim that another organ than the stomach—namely, the pancreas—has a share in digestion, and in the course of the ensuing decade it came to be known, through the efforts of Eberle, Valentin, and Claude Bernard, that this organ is all—important in the digestion of starchy and fatty foods. It was found, too, that the liver and the intestinal glands have each an important share in the work of preparing foods for absorption, as also has the saliva—that, in short, a coalition of forces is necessary for the digestion of all ordinary foods taken into the stomach.

And the chemists soon discovered that in each one of the essential digestive juices there is at least one substance having certain resemblances to pepsin, though acting on different kinds of food. The point of resemblance between all these essential digestive agents is that each has the remarkable property of acting on relatively enormous quantities of the substance which it can digest without itself being destroyed or apparently even altered. In virtue of this strange property, pepsin and the allied substances were spoken of as ferments, but more recently it is customary to distinguish them from such organized ferments as yeast by designating them enzymes. The isolation of these enzymes, and an appreciation of their mode of action, mark a long step towards the solution of the riddle of digestion, but it must be added that we are still quite in the dark as to the real ultimate nature of their strange activity.

In a comprehensive view, the digestive organs, taken as a whole, are a gateway between the outside world and the more intimate cells of the organism. Another equally important gateway is furnished by the lungs, and here also there was much obscurity about the exact method of functioning at the time of the revival of physiological chemistry. That oxygen is consumed and carbonic acid given off during respiration the chemists of the age of Priestley and Lavoisier had indeed made clear, but the mistaken notion prevailed that it was in the lungs themselves that the important burning of fuel occurs, of which carbonic acid is a chief product. But now that attention had been called to the importance of the ultimate cell, this misconception could not long hold its ground, and as early as 1842 Liebig, in the course of his studies of animal heat, became convinced that it is not in the lungs, but in the ultimate tissues to which they are tributary, that the true consumption of fuel takes place. Reviving Lavoisier's idea, with modifications and additions, Liebig contended, and in the face of opposition finally demonstrated, that the source of animal heat is really the consumption of the fuel taken in through the stomach and the lungs. He showed that all the activities of life are really the product of energy liberated solely through destructive processes, amounting, broadly speaking, to combustion occurring in the ultimate cells of the organism. Here is his argument:

LIEBIG ON ANIMAL HEAT

"The oxygen taken into the system is taken out again in the same forms, whether in summer or in winter; hence we expire more carbon in cold weather, and when the barometer is high, than we do in warm weather; and we must consume more or less carbon in our food in the same proportion; in Sweden more than in Sicily; and in our more temperate climate a full eighth more in winter than in summer.

"Even when we consume equal weights of food in cold and warm countries, infinite wisdom has so arranged that the articles of food in different climates are most unequal in the proportion of carbon they contain. The fruits on which the natives of the South prefer to feed do not in the fresh state contain more than twelve per cent. of carbon, while the blubber and train—oil used by the inhabitants of the arctic regions contain from sixty—six to eighty per cent. of carbon.

"It is no difficult matter, in warm climates, to study moderation in eating, and men can bear hunger for a long time under the equator; but cold and hunger united very soon exhaust the body.

"The mutual action between the elements of the food and the oxygen conveyed by the circulation of the blood to every part of the body is the source of animal heat.

"All living creatures whose existence depends on the absorption of oxygen possess within themselves a source of heat independent of surrounding objects.

"This truth applies to all animals, and extends besides to the germination of seeds, to the flowering of plants, and to the maturation of fruits. It is only in those parts of the body to which arterial blood, and with it the oxygen absorbed in respiration, is conveyed that heat is produced. Hair, wool, or feathers do not possess an elevated temperature. This high temperature of the animal body, or, as it may be called, disengagement of heat, is uniformly and under all circumstances the result of the combination of combustible substance with oxygen.

"In whatever way carbon may combine with oxygen, the act of combination cannot take place without the disengagement of heat. It is a matter of indifference whether the combination takes place rapidly or slowly, at a high or at a low temperature; the amount of heat liberated is a constant quantity. The carbon of the food, which is converted into carbonic acid within the body, must give out exactly as much heat as if it had been directly burned in the air or in oxygen gas; the only difference is that the amount of heat produced is diffused over unequal times. In oxygen the combustion is more rapid and the heat more intense; in air it is slower, the temperature is not so high, but it continues longer.

"It is obvious that the amount of heat liberated must increase or diminish with the amount of oxygen introduced in equal times by respiration. Those animals which respire frequently, and consequently consume much oxygen, possess a higher temperature than others which, with a body of equal size to be heated, take into the system less oxygen. The temperature of a child (102 degrees) is higher than that of an adult (99.5 degrees). That of birds (104 to 105.4 degrees) is higher than that of quadrupeds (98.5 to 100.4

degrees), or than that of fishes or amphibia, whose proper temperature is from 3.7 to 2.6 degrees higher than that of the medium in which they live. All animals, strictly speaking, are warm–blooded; but in those only which possess lungs is the temperature of the body independent of the surrounding medium.

"The most trustworthy observations prove that in all climates, in the temperate zones as well as at the equator or the poles, the temperature of the body in man, and of what are commonly called warm–blooded animals, is invariably the same; yet how different are the circumstances in which they live.

"The animal body is a heated mass, which bears the same relation to surrounding objects as any other heated mass. It receives heat when the surrounding objects are hotter, it loses heat when they are colder than itself. We know that the rapidity of cooling increases with the difference between the heated body and that of the surrounding medium—that is, the colder the surrounding medium the shorter the time required for the cooling of the heated body. How unequal, then, must be the loss of heat of a man at Palermo, where the actual temperature is nearly equal to that of the body, and in the polar regions, where the external temperature is from 70 to 90 degrees lower.

"Yet notwithstanding this extremely unequal loss of heat, experience has shown that the blood of an inhabitant of the arctic circle has a temperature as high as that of the native of the South, who lives in so different a medium. This fact, when its true significance is perceived, proves that the heat given off to the surrounding medium is restored within the body with great rapidity. This compensation takes place more rapidly in winter than in summer, at the pole than at the equator.

"Now in different climates the quantity of oxygen introduced into the system of respiration, as has been already shown, varies according to the temperature of the external air; the quantity of inspired oxygen increases with the loss of heat by external cooling, and the quantity of carbon or hydrogen necessary to combine with this oxygen must be increased in like ratio. It is evident that the supply of heat lost by cooling is effected by the mutual action of the elements of the food and the inspired oxygen, which combine together. To make use of a familiar, but not on that account a less just illustration, the animal body acts, in this respect, as a furnace, which we supply with fuel. It signifies nothing what intermediate forms food may assume, what changes it may undergo in the body, the last change is uniformly the conversion of carbon into carbonic acid and of its hydrogen into water; the unassimilated nitrogen of the food, along with the unburned or unoxidized carbon, is expelled in the excretions. In order to keep up in a furnace a constant temperature, we must vary the supply of fuel according to the external temperature—that is, according to the supply of oxygen.

"In the animal body the food is the fuel; with a proper supply of oxygen we obtain the heat given out during its oxidation or combustion."[3]

BLOOD CORPUSCLES, MUSCLES, AND GLANDS

Further researches showed that the carriers of oxygen, from the time of its absorption in the lungs till its liberation in the ultimate tissues, are the red corpuscles, whose function had been supposed to be the mechanical one of mixing of the blood. It transpired that the red corpuscles are composed chiefly of a substance which Kuhne first isolated in crystalline form in 1865, and which was named haemoglobin—a substance which has a marvellous affinity for oxygen, seizing on it eagerly at the lungs yet giving it up with equal readiness when coursing among the remote cells of the body. When freighted with oxygen it becomes oxyhaemoglobin and is red in color; when freed from its oxygen it takes a purple hue; hence the widely different appearance of arterial and venous blood, which so puzzled the early physiologists.

This proof of the vitally important role played by the red–blood corpuscles led, naturally, to renewed studies of these infinitesimal bodies. It was found that they may vary greatly in number at different periods in the life of the same individual, proving that they may be both developed and destroyed in the adult organism. Indeed, extended observations left no reason to doubt that the process of corpuscle formation and destruction may be a perfectly normal one—that, in short, every red–blood corpuscle runs its course and dies like any more elaborate organism. They are formed constantly in the red marrow of bones, and are destroyed in the liver, where they contribute to the formation of the coloring matter of the bile. Whether there are other seats of such manufacture and destruction of the corpuscles is not yet fully determined. Nor are histologists agreed as to whether the red–blood corpuscles themselves are to be regarded as true cells, or merely as fragments of cells budded out from a true cell for a special purpose; but in either case there is not the slightest doubt that the chief function of the red corpuscle is to carry oxygen.

If the oxygen is taken to the ultimate cells before combining with the combustibles it is to consume, it goes without saying that these combustibles themselves must be carried there also. Nor could it be in doubt that the chiefest of these ultimate tissues, as regards, quantity of fuel required, are the muscles. A general and comprehensive view of the organism includes, then, digestive apparatus and lungs as the channels of fuel−supply; blood and lymph channels as the transportation system; and muscle cells, united into muscle fibres, as the consumption furnaces, where fuel is burned and energy transformed and rendered available for the purposes of the organism, supplemented by a set of excretory organs, through which the waste products—the ashes—are eliminated from the system.

But there remain, broadly speaking, two other sets of organs whose size demonstrates their importance in the economy of the organism, yet whose functions are not accounted for in this synopsis. These are those glandlike organs, such as the spleen, which have no ducts and produce no visible secretions, and the nervous mechanism, whose central organs are the brain and spinal cord. What offices do these sets of organs perform in the great labor−specializing aggregation of cells which we call a living organism?

As regards the ductless glands, the first clew to their function was given when the great Frenchman Claude Bernard (the man of whom his admirers loved to say, "He is not a physiologist merely; he is physiology itself") discovered what is spoken of as the glycogenic function of the liver. The liver itself, indeed, is not a ductless organ, but the quantity of its biliary output seems utterly disproportionate to its enormous size, particularly when it is considered that in the case of the human species the liver contains normally about one−fifth of all the blood in the entire body. Bernard discovered that the blood undergoes a change of composition in passing through the liver. The liver cells (the peculiar forms of which had been described by Purkinje, Henle, and Dutrochet about 1838) have the power to convert certain of the substances that come to them into a starchlike compound called glycogen, and to store this substance away till it is needed by the organism. This capacity of the liver cells is quite independent of the bile−making power of the same cells; hence the discovery of this glycogenic function showed that an organ may have more than one pronounced and important specific function. But its chief importance was in giving a clew to those intermediate processes between digestion and final assimilation that are now known to be of such vital significance in the economy of the organism.

In the forty odd years that have elapsed since this pioneer observation of Bernard, numerous facts have come to light showing the extreme importance of such intermediate alterations of food–supplies in the blood as that performed by the liver. It has been shown that the pancreas, the spleen, the thyroid gland, the suprarenal capsules are absolutely essential, each in its own way, to the health of the organism, through metabolic changes which they alone seem capable of performing; and it is suspected that various other tissues, including even the muscles themselves, have somewhat similar metabolic capacities in addition to their recognized functions. But so extremely intricate is the chemistry of the substances involved that in no single case has the exact nature of the metabolisms wrought by these organs been fully made out. Each is in its way a chemical laboratory indispensable to the right conduct of the organism, but the precise nature of its operations remains inscrutable. The vast importance of the operations of these intermediate organs is unquestioned.

A consideration of the functions of that other set of organs known collectively as the nervous system is reserved for a later chapter.

VI. THEORIES OF ORGANIC EVOLUTION

GOETHE AND THE METAMORPHOSIS OF PARTS

When Coleridge said of Humphry Davy that he might have been the greatest poet of his time had he not chosen rather to be the greatest chemist, it is possible that the enthusiasm of the friend outweighed the caution of the critic. But however that may be, it is beyond dispute that the man who actually was the greatest poet of that time might easily have taken the very highest rank as a scientist had not the muse distracted his attention. Indeed, despite these distractions, Johann Wolfgang von Goethe achieved successes in the field of pure science that would insure permanent recognition for his name had he never written a stanza of poetry. Such is the versatility that marks the highest genius.

It was in 1790 that Goethe published the work that laid the foundations of his scientific reputation—the work on the Metamorphoses of Plants, in which he advanced the novel doctrine that all parts of the flower are modified or metamorphosed leaves.

"Every one who observes the growth of plants, even superficially," wrote Goethe, "will notice that certain external parts of them become transformed at times and go over into the forms of the contiguous parts, now completely, now to a greater or less degree. Thus, for example, the single flower is transformed into a double one when, instead of stamens, petals are developed, which are either exactly like the other petals of the corolla in form, and color or else still bear visible signs of their origin.

"When we observe that it is possible for a plant in this way to take a step backward, we shall give so much the more heed to the regular course of nature and learn the laws of transformation according to which she produces one part through another, and displays the most varying forms through the modification of one single organ.

"Let us first direct our attention to the plant at the moment when it develops out of the seed–kernel. The first organs of its upward growth are known by the name of cotyledons; they have also been called seed–leaves.

"They often appear shapeless, filled with new matter, and are just as thick as they are broad. Their vessels are unrecognizable and are hardly to be distinguished from the mass of the whole; they bear almost no resemblance to a leaf, and we could easily be misled into regarding them as special organs. Occasionally, however, they appear as real leaves, their vessels are capable of the most minute development, their similarity to the following leaves does not permit us to take them for special organs, but we recognize them instead to be the first leaves of the stalk.

"The cotyledons are mostly double, and there is an observation to be made here which will appear still more important as we proceed—that is, that the leaves of the first node are often paired, even when the following leaves of the stalk stand alternately upon it. Here we see an approximation and a joining of parts which nature afterwards separates and places at a distance from one another. It is still more remarkable when the cotyledons take the form of many little leaves gathered about an axis, and the stalk which grows gradually from their midst produces the following leaves arranged around it singly in a whorl. This may be observed very exactly in the growth of the pinus species. Here a corolla of needles forms at the same time a calyx, and we shall have occasion to remember the present case in connection with similar phenomena later.

"On the other hand, we observe that even the cotyledons which are most like a leaf when compared with the following leaves of the stalk are always more undeveloped or less developed. This is chiefly noticeable in their margin which is extremely simple and shows few traces of indentation.

"A few or many of the next following leaves are often already present in the seed, and lie enclosed between the cotyledons; in their folded state they are known by the name of plumules. Their form, as compared with the cotyledons and the following leaves, varies in different plants. Their chief point of variance, however, from the cotyledons is that they are flat, delicate, and formed like real leaves generally. They are wholly green, rest on a visible node, and can no longer deny their relationship to the following leaves of the stalk, to which, however, they are usually still inferior, in so far as that their margin is not completely developed.

"The further development, however, goes on ceaselessly in the leaf, from node to node; its midrib is elongated, and more or less additional ribs stretch out from this towards the sides. The leaves now appear notched, deeply indented, or composed of several small leaves, in which last case they seem to form complete little branches. The date–palm furnishes a striking example of such a successive transformation of the simplest leaf form. A midrib is elongated through a succession of several leaves, the single fan–shaped leaf becomes torn and diverted, and a very complicated leaf is developed, which rivals a branch in form.

"The transition to inflorescence takes place more or less rapidly. In the latter case we usually observe that the leaves of the stalk loose their different external divisions, and, on the other hand, spread out more or less in their lower parts where they are attached to the stalk. If the transition takes place rapidly, the stalk, suddenly become thinner and more elongated since the node of the last–developed leaf, shoots up and collects several leaves around an axis at its end.

"That the petals of the calyx are precisely the same organs which have hitherto appeared as leaves on the stalk, but now stand grouped about a common centre in an often very different form, can, as it seems to me, be most clearly demonstrated. Already in connection with the cotyledons above, we noticed a similar working of nature. The first species, while they are developing out of the seed–kernel, display a radiate crown of unmistakable needles; and in the first childhood of these plants we see already indicated that force of nature whereby when they are older their flowering and fruit–giving state will be produced.

"We see this force of nature, which collects several leaves around an axis, produce a still closer union and make these approximated, modified leaves still more unrecognizable by joining them together either wholly or partially. The bell–shaped or so–called one–petalled calices represent these cloudy connected leaves, which, being more or less indented from above, or divided, plainly show their origin.

"We can observe the transition from the calyx to the corolla in more than one instance, for, although the color of the calyx is still usually green, and like the color of the leaves of the stalk, it nevertheless often varies in one or another of its parts—at the tips, the margins, the back, or even, the inward side—while the outer still remains on green.

"The relationship of the corolla to the leaves of the stalk is shown in more than one way, since on the stalks of some plants appear leaves which are already more or less colored long before they approach inflorescence; others are fully colored when near inflorescence. Nature also goes over at once to the corolla, sometimes by skipping over the organs of the calyx, and in such a case we likewise have an opportunity to observe that leaves of the stalk become transformed into petals. Thus on the stalk of tulips, for instance, there sometimes appears an almost completely developed and colored petal. Even more remarkable is the case when such a leaf, half green and half of it belonging to the stalk, remains attached to the latter, while another colored part is raised with the corolla, and the leaf is thus torn in two.

"The relationship between the petals and stamens is very close. In some instances nature makes the transition regular—e.g., among the Canna and several plants of the same family. A true, little–modified petal is drawn together on its upper margin, and produces a pollen sac, while the rest of the petal takes the place of the stamen. In double flowers we can observe this transition in all its stages. In several kinds of roses, within the fully developed and colored petals there appear other ones which are drawn together in the middle or on the side. This drawing together is produced by a small weal, which appears as a more or less complete pollen sac, and in the same proportion the leaf approaches the simple form of a stamen.

"The pistil in many cases looks almost like a stamen without anthers, and the relationship between the formation of the two is much closer than between the other parts. In retrograde fashion nature often produces cases where the style and stigma (Narben) become retransformed into petals—that is, the Ranunculus Asiaticus becomes double by transforming the stigma and style of the fruit–receptacle into real petals, while the stamens are often found unchanged immediately behind the corolla.

"In the seed receptacles, in spite of their formation, of their special object, and of their method of being joined together, we cannot fail to recognize the leaf form. Thus, for instance, the pod would be a simple leaf folded and grown together on its margin; the siliqua would consist of more leaves folded over another; the compound receptacles would be explained as being several leaves which, being united above one centre, keep their inward parts separate and are joined on their margins. We can convince ourselves of this by actual sight when such composite capsules fall apart after becoming ripe, because then every part displays an opened pod."[1]

The theory thus elaborated of the metamorphosis of parts was presently given greater generality through extension to the animal kingdom, in the doctrine which Goethe and Oken advanced independently, that the vertebrate skull is essentially a modified and developed vertebra. These were conceptions worthy of a poet—impossible, indeed, for any mind that had not the poetic faculty of correlation. But in this case the poet's vision was prophetic of a future view of the most prosaic science. The doctrine of metamorphosis of parts soon came to be regarded as of fundamental importance.

But the doctrine had implications that few of its early advocates realized. If all the parts of a flower—sepal, petal, stamen, pistil, with their countless deviations of contour and color—are but modifications of the leaf, such modification implies a marvellous differentiation and development. To assert that a stamen is a metamorphosed leaf means, if it means anything, that in the long sweep of time the leaf has by slow or sudden gradations changed its character through successive generations, until the offspring, so to speak, of a true leaf has become a stamen. But if such a metamorphosis as this is possible—if the seemingly wide gap between leaf and stamen may be spanned by the modification of a line of organisms—where does the possibility of modification of organic type find its bounds? Why may not the modification of parts go on along devious lines until the remote descendants of an organism are utterly unlike that organism? Why may we not thus account for the development of various species of beings all sprung from one parent stock? That, too, is a poet's dream; but is it only a dream? Goethe thought not. Out of his studies of metamorphosis of parts there grew in his mind the belief that the multitudinous species of plants and animals about us have been evolved from fewer and fewer earlier parent types, like twigs of a giant tree drawing their nurture from the same primal root. It was a bold and revolutionary thought, and the world regarded it as but the vagary of a poet.

ERASMUS DARWIN

Just at the time when this thought was taking form in Goethe's brain, the same idea was germinating in the mind of another philosopher, an Englishman of international fame, Dr. Erasmus Darwin, who, while he lived, enjoyed the widest popularity as a poet, the rhymed couplets of his Botanic Garden being quoted everywhere with admiration. And posterity repudiating the verse which makes the body of the book, yet grants permanent value to the book itself, because, forsooth, its copious explanatory foot-notes furnish an outline of the status of almost every department of science of the time.

But even though he lacked the highest art of the versifier, Darwin had, beyond peradventure, the imagination of a poet coupled with profound scientific knowledge; and it was his poetic insight, correlating organisms seemingly diverse in structure and imbuing the lowliest flower with a vital personality, which led him to suspect that there are no lines of demarcation in nature. "Can it be," he queries, "that one form of organism has developed from another; that different species are really but modified descendants of one parent stock?" The alluring thought nestled in his mind and was nurtured there, and grew in a fixed belief, which was given fuller expression in his Zoonomia and in the posthumous Temple of Nature.

Here is his rendering of the idea as versified in the Temple of Nature:

"Organic life beneath the shoreless waves
 Was born, and nursed in Ocean's pearly caves;
 First forms minute, unseen by spheric glass,
 Move on the mud, or pierce the watery mass;
 These, as successive generations bloom,
 New powers acquire and larger limbs assume;
 Whence countless groups of vegetation spring,
 And breathing realms of fin, and feet, and wing.

"Thus the tall Oak, the giant of the wood,
 Which bears Britannia's thunders on the flood;

The Whale, unmeasured monster of the main;
The lordly lion, monarch of the plain;
The eagle, soaring in the realms of air,
Whose eye, undazzled, drinks the solar glare;
Imperious man, who rules the bestial crowd,
Of language, reason, and reflection proud,
With brow erect, who scorns this earthy sod,
And styles himself the image of his God—
Arose from rudiments of form and sense,
An embryon point or microscopic ens!"[2]

Here, clearly enough, is the idea of evolution. But in that day there was little proof forthcoming of its validity that could satisfy any one but a poet, and when Erasmus Darwin died, in 1802, the idea of transmutation of species was still but an unsubstantiated dream.

It was a dream, however, which was not confined to Goethe and Darwin. Even earlier the idea had come more or less vaguely to another great dreamer—and worker—of Germany, Immanuel Kant, and to several great Frenchmen, including De Maillet, Maupertuis, Robinet, and the famous naturalist Buffon—a man who had the imagination of a poet, though his message was couched in most artistic prose. Not long after the middle of the eighteenth century Buffon had put forward the idea of transmutation of species, and he reiterated it from time to time from then on till his death in 1788. But the time was not yet ripe for the idea of transmutation of species to burst its bonds.

And yet this idea, in a modified or undeveloped form, had taken strange hold upon the generation that was upon the scene at the close of the eighteenth century. Vast numbers of hitherto unknown species of animals had been recently discovered in previously unexplored regions of the globe, and the wise men were sorely puzzled to account for the disposal of all of these at the time of the deluge. It simplified matters greatly to suppose that many existing species had been developed since the episode of the ark by modification of the original pairs. The remoter bearings of such a theory were overlooked for the time, and the idea that American animals and birds, for example, were modified descendants of Old–World forms—the jaguar of the leopard, the puma of the lion, and so on—became a current belief with that class of humanity who accept almost any statement as true that harmonizes with their prejudices without realizing its implications.

Thus it is recorded with eclat that the discovery of the close proximity of America at the northwest with Asia removes all difficulties as to the origin of the Occidental faunas and floras, since Oriental species might easily have found their way to America on the ice, and have been modified as we find them by "the well–known influence of climate." And the persons who gave expression to this idea never dreamed of its real significance. In truth, here was the doctrine of evolution in a nutshell, and, because its ultimate bearings were not clear, it seemed the most natural of doctrines. But most of the persons who advanced it would have turned from it aghast could they have realized its import. As it was, however, only here and there a man like Buffon reasoned far enough to inquire what might be the limits of such assumed transmutation; and only here and there a Darwin or a Goethe reached the conviction that there are no limits.

LAMARCK VERSUS CUVIER

And even Goethe and Darwin had scarcely passed beyond that tentative stage of conviction in which they held the thought of transmutation of species as an ancillary belief not ready for full exposition. There was one of their contemporaries, however, who, holding the same conception, was moved to give it full explication. This was the friend and disciple of Buffon, Jean Baptiste de Lamarck. Possessed of the spirit of a poet and philosopher, this great Frenchman had also the widest range of technical knowledge, covering the entire field of animate nature. The first half of his long life was devoted chiefly to botany, in which he attained high distinction. Then, just at the beginning of the nineteenth century, he turned to zoology, in particular to the lower forms of animal life. Studying these lowly organisms, existing and fossil, he was more and more impressed with the gradations of form everywhere to be seen; the linking of diverse families through intermediate ones; and in particular with the predominance of low types of life in the earlier geological strata. Called upon constantly to classify the various forms of life in the course of his systematic writings, he found it more and more difficult to draw sharp lines of demarcation, and at last the suspicion long harbored grew into a settled conviction that there is really no such thing as a species of organism in nature; that "species" is a figment of the human imagination, whereas in nature there are only individuals.

That certain sets of individuals are more like one another than like other sets is of course patent, but this only means, said Lamarck, that these similar groups have had comparatively recent common ancestors, while dissimilar sets of beings are more remotely related in consanguinity. But trace back the lines of descent far enough, and all will culminate in one original stock. All forms of life whatsoever are modified descendants of an original organism. From lowest to highest, then, there is but one race, one species, just as all the multitudinous branches and twigs from one root are but one tree. For purposes of convenience of description, we may divide organisms into orders, families, genera, species, just as we divide a tree into root, trunk, branches, twigs, leaves; but in the one case, as in the other, the division is arbitrary and artificial.

In Philosophie Zoologique (1809), Lamarck first explicitly formulated his ideas as to the transmutation of species, though he had outlined them as early as 1801. In this memorable publication not only did he state his belief more explicitly and in fuller detail than the idea had been expressed by any predecessor, but he took another long forward step, carrying him far beyond all his forerunners except Darwin, in that he made an attempt to explain the way in which the transmutation of species had been brought about. The changes have been wrought, he said, through the unceasing efforts of each organism to meet the needs imposed upon it by its environment. Constant striving means the constant use of certain organs. Thus a bird running by the seashore is constantly tempted to wade deeper and deeper in pursuit of food; its incessant efforts tend to develop its legs, in accordance with the observed principle that the use of any organ tends to strengthen and develop it. But such slightly increased development of the legs is transmitted to the off spring of the bird, which in turn develops its already improved legs by its individual efforts, and transmits the improved tendency. Generation after generation this is repeated, until the sum of the infinitesimal variations, all in the same direction, results in the production of the long–legged wading–bird. In a similar way, through individual effort and transmitted tendency, all the diversified organs of all creatures have been developed—the fin of the fish, the wing of the bird, the hand of man; nay, more, the fish itself, the bird, the man, even. Collectively the organs make up the entire organism; and what is true of the individual organs must be true also of their ensemble, the living being.

Whatever might be thought of Lamarck's explanation of the cause of transmutation—which really was that already suggested by Erasmus Darwin—the idea of the evolution for which he contended was but the logical extension of the conception that American animals are the modified and degenerated descendants of European animals. But people as a rule are little prone to follow ideas to their logical conclusions, and in this case the conclusions were so utterly opposed to the proximal bearings of the idea that the whole thinking world repudiated them with acclaim. The very persons who had most eagerly accepted the idea of transmutation of European species into American species, and similar limited variations through changed environment, because of the relief thus given the otherwise overcrowded ark, were now foremost in denouncing such an extension of the doctrine of transmutation as Lamarck proposed.

And, for that matter, the leaders of the scientific world were equally antagonistic to the Lamarckian hypothesis. Cuvier in particular, once the pupil of Lamarck, but now his colleague, and in authority more than his peer, stood out against the transmutation doctrine with all his force. He argued for the absolute fixity of species, bringing to bear the resources of a mind which, as a mere repository of facts, perhaps never was excelled. As a final and tangible proof of his position, he brought forward the bodies of ibises that had been embalmed by the ancient Egyptians, and showed by comparison that these do not differ in the slightest particular from the ibises that visit the Nile to–day.

Cuvier's reasoning has such great historical interest—being the argument of the greatest opponent of evolution of that day—that we quote it at some length.

"The following objections," he says, "have already been started against my conclusions. Why may not the presently existing races of mammiferous land quadrupeds be mere modifications or varieties of those ancient races which we now find in the fossil state, which modifications may have been produced by change of climate and other local circumstances, and since raised to the present excessive difference by the operations of similar causes during a long period of ages?

"This objection may appear strong to those who believe in the indefinite possibility of change of form in organized bodies, and think that, during a succession of ages and by alterations of habitudes, all the species may change into one another, or one of them give birth to all the rest. Yet to these persons the following answer may be given from their own system: If the species have changed by degrees, as they assume, we ought to find traces of this gradual modification. Thus, between the palaeotherium and the species of our own day, we should be able to discover some intermediate forms; and yet no such discovery has ever been made. Since the bowels of the earth have not preserved monuments of this strange genealogy, we have no right to conclude that the ancient and now extinct species were as permanent in their forms and characters as those which exist at present; or, at least, that the catastrophe which destroyed them did not leave sufficient time for the productions of the changes that are alleged to have taken place.

"In order to reply to those naturalists who acknowledge that the varieties of animals are restrained by nature within certain limits, it would be necessary to examine how far these limits extend. This is a very curious inquiry, and in itself exceedingly interesting under a variety of relations, but has been hitherto very little attended to. . . . .

Wild animals which subsist upon herbage feel the influence of climate a little more extensively, because there is added to it the influence of food, both in regard to its abundance and its quality. Thus the elephants of one forest are larger than those of another; their tusks also grow somewhat longer in places where their food may happen to be more favorable for the production of the substance of ivory. The same may take place in regard to the horns of stags and reindeer. But let us examine two elephants, the most dissimilar that can be conceived, we shall not discover the smallest difference in the number and articulations of the bones, the structure of the teeth, etc. . . . .

"Nature appears also to have guarded against the alterations of species which might proceed from mixture of breeds by influencing the various species of animals with mutual aversion from one another. Hence all the cunning and all the force that man is able to exert is necessary to accomplish such unions, even between species that have the nearest resemblances. And when the mule breeds that are thus produced by these forced conjunctions happen to be fruitful, which is seldom the case, this fecundity never continues beyond a few generations, and would not probably proceed so far without a continuance of the same cares which excited it at first. Thus we never see in a wild state intermediate productions between the hare and the rabbit, between the stag and the doe, or between the marten and the weasel. But the power of man changes this established order, and continues to produce all these intermixtures of which the various species are susceptible, but which they would never produce if left to themselves.

"The degrees of these variations are proportional to the intensity of the causes that produced them—namely, the slavery or subjection under which those animals are to man. They do not proceed far in half-domesticated species. In the cat, for example, a softer or harsher fur, more brilliant or more varied colors, greater or less size—these form the whole extent of variety in the species; the skeleton of the cat of Angora differs in no regular and constant circumstances from the wild-cat of Europe. . . . .

The most remarkable effects of the influence of man are produced upon that animal which he has reduced most completely under subjection. Dogs have been transported by mankind into every part of the world and have submitted their action to his entire direction. Regulated in their unions by the pleasure or caprice of their masters, the almost endless varieties of dogs differ from one another in color, in length, and abundance of hair, which is sometimes entirely wanting; in their natural instincts; in size, which varies in measure as one to five, mounting in some instances to more than a hundredfold in bulk; in the form of their ears, noses, and tails; in the relative length of their legs; in the progressive development of the brain, in several of the domesticated varieties occasioning alterations even in the form of the head, some of them having long, slender muzzles with a flat forehead, others having short muzzles with a forehead convex, etc., insomuch that the apparent difference between a mastiff and a water–spaniel and between a greyhound and a pugdog are even more striking than between almost any of the wild species of a genus. . . . .

It follows from these observations that animals have certain fixed and natural characters which resist the effects of every kind of influence, whether proceeding from natural causes or human interference; and we have not the smallest reason to suspect that time has any more effect on them than climate.

"I am aware that some naturalists lay prodigious stress upon the thousands which they can call into action by a dash of their pens. In such matters, however, our only way of judging as to the effects which may be produced by a long period of time is by multiplying, as it were, such as are produced by a shorter time. With this view I have endeavored to collect all the ancient documents respecting the forms of animals; and there are none equal to those furnished by the Egyptians, both in regard to their antiquity and abundance. They have not only left us representatives of animals, but even their identical bodies embalmed and preserved in the catacombs.

"I have examined, with the greatest attention, the engraved figures of quadrupeds and birds brought from Egypt to ancient Rome, and all these figures, one with another, have a perfect resemblance to their intended objects, such as they still are to–day.

"From all these established facts, there does not seem to be the smallest foundation for supposing that the new genera which I have discovered or established among extraneous fossils, such as the paleoetherium, anoplotherium, megalonyx, mastodon, pterodactylis, etc., have ever been the sources of any of our present animals, which only differ so far as they are influenced by time or climate. Even if it should prove true, which I am far from believing to be the case, that the fossil elephants, rhinoceroses, elks, and bears do not differ further from the existing species of the same genera than the present races of dogs differ among themselves, this would by no means be a sufficient reason to conclude that they were of the same species; since the races or varieties of dogs have been influenced by the trammels of domesticity, which those other animals never did, and indeed never could, experience."[3]

To Cuvier's argument from the fixity of Egyptian mummified birds and animals, as above stated, Lamarck replied that this proved nothing except that the ibis had become perfectly adapted to its Egyptian surroundings in an early day, historically speaking, and that the climatic and other conditions of the Nile Valley had not since then changed. His theory, he alleged, provided for the stability of species under fixed conditions quite as well as for transmutation under varying conditions.

But, needless to say, the popular verdict lay with Cuvier; talent won for the time against genius, and Lamarck was looked upon as an impious visionary. His faith never wavered, however. He believed that he had gained a true insight into the processes of animate nature, and he reiterated his hypotheses over and over, particularly in the introduction to his Histoire Naturelle des Animaux sans Vertebres, in 1815, and in his Systeme des Connaissances Positives de l'Homme, in 1820. He lived on till 1829, respected as a naturalist, but almost unrecognized as a prophet.

TENTATIVE ADVANCES

While the names of Darwin and Goethe, and in particular that of Lamarck, must always stand out in high relief in this generation as the exponents of the idea of transmutation of species, there are a few others which must not be altogether overlooked in this connection. Of these the most conspicuous is that of Gottfried Reinhold Treviranus, a German naturalist physician, professor of mathematics in the lyceum at Bremen.

It was an interesting coincidence that Treviranus should have published the first volume of his Biologie, oder Philosophie der lebenden Natur, in which his views on the transmutation of species were expounded, in 1802, the same twelvemonth in which Lamarck's first exposition of the same doctrine appeared in his Recherches sur l'Organisation des Corps Vivants. It is singular, too, that Lamarck, in his Hydrogelogie of the same date, should independently have suggested "biology" as an appropriate word to express the general science of living things. It is significant of the tendency of thought of the time that the need of such a unifying word should have presented itself simultaneously to independent thinkers in different countries.

That same memorable year, Lorenz Oken, another philosophical naturalist, professor in the University of Zurich, published the preliminary outlines of his Philosophie der Natur, which, as developed through later publications, outlined a theory of spontaneous generation and of evolution of species. Thus it appears that this idea was germinating in the minds of several of the ablest men of the time during the first decade of our century. But the singular result of their various explications was to give sudden check to that undercurrent of thought which for some time had been setting towards this conception. As soon as it was made clear whither the concession that animals may be changed by their environment must logically trend, the recoil from the idea was instantaneous and fervid. Then for a generation Cuvier was almost absolutely dominant, and his verdict was generally considered final.

There was, indeed, one naturalist of authority in France who had the hardihood to stand out against Cuvier and his school, and who was in a position to gain a hearing, though by no means to divide the following. This was Etienne Geoffroy Saint-Hilaire, the famous author of the Philosophie Anatomique, and for many years the colleague of Lamarck at the Jardin des Plantes. Like Goethe, Geoffroy was pre-eminently an anatomist, and, like the great German, he had early been impressed with the resemblances between the analogous organs of different classes of beings. He conceived the idea that an absolute unity of type prevails throughout organic nature as regards each set of organs. Out of this idea grew his gradually formed belief that similarity of structure might imply identity of origin—that, in short, one species of animal might have developed from another.

Geoffroy's grasp of this idea of transmutation was by no means so complete as that of Lamarck, and he seems never to have fully determined in his own mind just what might be the limits of such development of species. Certainly he nowhere includes all organic creatures in one line of descent, as Lamarck had done; nevertheless, he held tenaciously to the truth as he saw it, in open opposition to Cuvier, with whom he held a memorable debate at the Academy of Sciences in 1830—the debate which so aroused the interest and enthusiasm of Goethe, but which, in the opinion of nearly every one else, resulted in crushing defeat for Geoffrey, and brilliant, seemingly final, victory for the advocate of special creation and the fixity of species.

With that all ardent controversy over the subject seemed to end, and for just a quarter of a century to come there was published but a single argument for transmutation of species which attracted any general attention whatever. This oasis in a desert generation was a little book called Vestiges of the Natural History of Creation, which appeared anonymously in England in 1844, and which passed through numerous editions, and was the subject of no end of abusive and derisive comment. This book, the authorship of which remained for forty years a secret, is now conceded to have been the work of Robert Chambers, the well-known English author and publisher. The book itself is remarkable as being an avowed and unequivocal exposition of a general doctrine of evolution, its view being as radical and comprehensive as that of Lamarck himself. But it was a resume of earlier efforts rather than a new departure, to say nothing of its technical shortcomings, which may best be illustrated by a quotation.

"The whole question," says Chambers, "stands thus: For the theory of universal order—that is, order as presiding in both the origin and administration of the world—we have the testimony of a vast number of facts in nature, and this one in addition—that whatever is left from the domain of ignorance, and made undoubted matter of science, forms a new support to the same doctrine. The opposite view, once predominant, has been shrinking for ages into lesser space, and now maintains a footing only in a few departments of nature which happen to be less liable than others to a clear investigation. The chief of these, if not almost the only one, is the origin of the organic kingdoms. So long as this remains obscure, the supernatural will have a certain hold upon enlightened persons. Should it ever be cleared up in a way that leaves no doubt of a natural origin of plants and animals, there must be a complete revolution in the view which is generally taken of the relation of the Father of our being.

"This prepares the way for a few remarks on the present state of opinion with regard to the origin of organic nature. The great difficulty here is the apparent determinateness of species. These forms of life being apparently unchangeable, or at least always showing a tendency to return to the character from which they have diverged, the idea arises that there can have been no progression from one to another; each must have taken its special form, independently of other forms, directly from the appointment of the Creator. The Edinburgh Review writer says, 'they were created by the hand of God and adapted to the conditions of the period.' Now it is, in the first place, not certain that species constantly maintain a fixed character, for we have seen that what were long considered as determinate species have been transmuted into others. Passing, however, from this fact, as it is not generally received among men of science, there remain some great difficulties in connection with the idea of special creation. First we should have to suppose, as pointed out in my former volume, a most startling diversity of plan in the divine workings, a great general plan or system of law in the leading events of world–making, and a plan of minute, nice operation, and special attention in some of the mere details of the process. The discrepancy between the two conceptions is surely overpowering, when we allow ourselves to see the whole matter in a steady and rational light. There is, also, the striking fact of an ascertained historical progress of plants and animals in the order of their organization; marine and cellular plants and invertebrated animals first, afterwards higher examples of both. In an arbitrary system we had surely no reason to expect mammals after reptiles; yet in this order they came. The writer in the Edinburgh Review speaks of animals as coming in adaptation to conditions, but this is only true in a limited sense. The groves which formed the coal–beds might have been a fitting habitation for reptiles, birds, and mammals, as such groves are at the present day; yet we see none of the last of these classes and hardly any traces of the two first at that period of the earth. Where the iguanodon lived the elephant might have lived, but there was no elephant at that time. The sea of the Lower Silurian era was capable of supporting fish, but no fish existed. It hence forcibly appears that theatres of life must have remained unserviceable, or in the possession of a tenantry inferior to what might have enjoyed them, for many ages: there surely would have been no such waste allowed in a system where Omnipotence was working upon the plan of minute attention to specialities. The fact seems to denote that the actual procedure of the peopling of the earth was one of a natural kind, requiring a long space of time for its evolution. In this supposition the long existence of land without land animals, and more particularly without the noblest classes and orders, is only analogous to the fact, not nearly enough present to the minds of a civilized people, that to this day the bulk of the earth is a waste as far as man is concerned.

"Another startling objection is in the infinite local variation of organic forms. Did the vegetable and animal kingdoms consist of a definite number of species adapted to peculiarities of soil and climate, and universally distributed, the fact would be in harmony with the idea of special exertion. But the truth is that various regions exhibit variations altogether without apparent end or purpose. Professor Henslow enumerates forty–five distinct flowers or sets of plants upon the surface of the earth, notwithstanding that many of these would be equally suitable elsewhere. The animals of different continents are equally various, few species being the same in any two, though the general character may conform. The inference at present drawn from this fact is that there must have been, to use the language of the Rev. Dr. Pye Smith, 'separate and original creations, perhaps at different and respectively distinct epochs.' It seems hardly conceivable that rational men should give an adherence to such a doctrine when we think of what it involves. In the single fact that it necessitates a special fiat of the inconceivable Author of this sand-cloud of worlds to produce the flora of St. Helena, we read its more than sufficient condemnation. It surely harmonizes far better with our general ideas of nature to suppose that, just as all else in this far-spread science was formed on the laws impressed upon it at first by its Author, so also was this. An exception presented to us in such a light appears admissible only when we succeed in forbidding our minds to follow out those reasoning processes to which, by another law of the Almighty, they tend, and for which they are adapted."[4]

Such reasoning as this naturally aroused bitter animadversions, and cannot have been without effect in creating an undercurrent of thought in opposition to the main trend of opinion of the time. But the book can hardly be said to have done more than that. Indeed, some critics have denied it even this merit. After its publication, as before, the conception of transmutation of species remained in the popular estimation, both lay and scientific, an almost forgotten "heresy."

It is true that here and there a scientist of greater or less repute—as Von Buch, Meckel, and Von Baer in Germany, Bory Saint-Vincent in France, Wells, Grant, and Matthew in England, and Leidy in America—had expressed more or less tentative dissent from the doctrine of special creation and immutability of species, but their unaggressive suggestions, usually put forward in obscure publications, and incidentally, were utterly overlooked and ignored. And so, despite the scientific advances along many lines at the middle of the century, the idea of the transmutability of organic races had no such prominence, either in scientific or unscientific circles, as it had acquired fifty years before. Special creation held the day, seemingly unopposed.

DARWIN AND THE ORIGIN OF SPECIES

But even at this time the fancied security of the special-creation hypothesis was by no means real. Though it seemed so invincible, its real position was that of an apparently impregnable fortress beneath which, all unbeknown to the garrison, a powder-mine has been dug and lies ready for explosion. For already there existed in the secluded work—room of an English naturalist, a manuscript volume and a portfolio of notes which might have sufficed, if given publicity, to shatter the entire structure of the special—creation hypothesis. The naturalist who, by dint of long and patient effort, had constructed this powder—mine of facts was Charles Robert Darwin, grandson of the author of Zoonomia.

As long ago as July 1, 1837, young Darwin, then twenty—eight years of age, had opened a private journal, in which he purposed to record all facts that came to him which seemed to have any bearing on the moot point of the doctrine of transmutation of species. Four or five years earlier, during the course of that famous trip around the world with Admiral Fitzroy, as naturalist to the Beagle, Darwin had made the personal observations which first tended to shake his belief of the fixity of species. In South America, in the Pampean formation, he had discovered "great fossil animals covered with armor like that on the existing armadillos," and had been struck with this similarity of type between ancient and existing faunas of the same region. He was also greatly impressed by the manner in which closely related species of animals were observed to replace one another as he proceeded southward over the continent; and "by the South—American character of most of the productions of the Galapagos Archipelago, and more especially by the manner in which they differ slightly on each island of the group, none of the islands appearing to be very ancient in a geological sense."

At first the full force of these observations did not strike him; for, under sway of Lyell's geological conceptions, he tentatively explained the relative absence of life on one of the Galapagos Islands by suggesting that perhaps no species had been created since that island arose. But gradually it dawned upon him that such facts as he had observed "could only be explained on the supposition that species gradually become modified." From then on, as he afterwards asserted, the subject haunted him; hence the journal of 1837.

It will thus be seen that the idea of the variability of species came to Charles Darwin as an inference from personal observations in the field, not as a thought borrowed from books. He had, of course, read the works of his grandfather much earlier in life, but the arguments of Zoonomia and The Temple of Nature had not served in the least to weaken his acceptance of the current belief in fixity of species. Nor had he been more impressed with the doctrine of Lamarck, so closely similar to that of his grandfather. Indeed, even after his South—American experience had aroused him to a new point of view he was still unable to see anything of value in these earlier attempts at an explanation of the variation of species. In opening his journal, therefore, he had no preconceived notion of upholding the views of these or any other makers of hypotheses, nor at the time had he formulated any hypothesis of his own. His mind was open and receptive; he was eager only for facts which might lead him to an understanding of a problem which seemed utterly obscure. It was something to feel sure that species have varied; but how have such variations been brought about?

It was not long before Darwin found a clew which he thought might lead to the answer he sought. In casting about for facts he had soon discovered that the most available field for observation lay among domesticated animals, whose numerous variations within specific lines are familiar to every one. Thus under domestication creatures so tangibly different as a mastiff and a terrier have sprung from a common stock. So have the Shetland pony, the thoroughbred, and the draught–horse. In short, there is no domesticated animal that has not developed varieties deviating more or less widely from the parent stock. Now, how has this been accomplished? Why, clearly, by the preservation, through selective breeding, of seemingly accidental variations. Thus one horseman, by constantly selecting animals that "chance" to have the right build and stamina, finally develops a race of running–horses; while another horseman, by selecting a different series of progenitors, has developed a race of slow, heavy draught animals.

So far, so good; the preservation of "accidental" variations through selective breeding is plainly a means by which races may be developed that are very different from their original parent form. But this is under man's supervision and direction. By what process could such selection be brought about among creatures in a state of nature? Here surely was a puzzle, and one that must be solved before another step could be taken in this direction.

The key to the solution of this puzzle came into Darwin's mind through a chance reading of the famous essay on "Population" which Thomas Robert Malthus had published almost half a century before. This essay, expositing ideas by no means exclusively original with Malthus, emphasizes the fact that organisms tend to increase at a geometrical ratio through successive generations, and hence would overpopulate the earth if not somehow kept in check. Cogitating this thought, Darwin gained a new insight into the processes of nature. He saw that in virtue of this tendency of each race of beings to overpopulate the earth, the entire organic world, animal and vegetable, must be in a state of perpetual carnage and strife, individual against individual, fighting for sustenance and life.

That idea fully imagined, it becomes plain that a selective influence is all the time at work in nature, since only a few individuals, relatively, of each generation can come to maturity, and these few must, naturally, be those best fitted to battle with the particular circumstances in the midst of which they are placed. In other words, the individuals best adapted to their surroundings will, on the average, be those that grow to maturity and produce offspring. To these offspring will be transmitted the favorable peculiarities. Thus these peculiarities will become permanent, and nature will have accomplished precisely what the human breeder is seen to accomplish. Grant that organisms in a state of nature vary, however slightly, one from another (which is indubitable), and that such variations will be transmitted by a parent to its offspring (which no one then doubted); grant, further, that there is incessant strife among the various organisms, so that only a small proportion can come to maturity—grant these things, said Darwin, and we have an explanation of the preservation of variations which leads on to the transmutation of species themselves.

This wonderful coign of vantage Darwin had reached by 1839. Here was the full outline of his theory; here were the ideas which afterwards came to be embalmed in familiar speech in the phrases "spontaneous variation," and the "survival of the fittest," through "natural selection." After such a discovery any ordinary man would at once have run through the streets of science, so to speak, screaming "Eureka!" Not so Darwin. He placed the manuscript outline of his theory in his portfolio, and went on gathering facts bearing on his discovery. In 1844 he made an abstract in a manuscript book of the mass of facts by that time accumulated. He showed it to his friend Hooker, made careful provision for its publication in the event of his sudden death, then stored it away in his desk and went ahead with the gathering of more data. This was the unexploded powder–mine to which I have just referred.

Twelve years more elapsed—years during which the silent worker gathered a prodigious mass of facts, answered a multitude of objections that arose in his own mind, vastly fortified his theory. All this time the toiler was an invalid, never knowing a day free from illness and discomfort, obliged to husband his strength, never able to work more than an hour and a half at a stretch; yet he accomplished what would have been vast achievements for half a dozen men of robust health. Two friends among the eminent scientists of the day knew of his labors—Sir Joseph Hooker, the botanist, and Sir Charles Lyell, the geologist. Gradually Hooker had come to be more than half a convert to Darwin's views. Lyell was still sceptical, yet he urged Darwin to publish his theory without further delay lest he be forestalled. At last the patient worker decided to comply with this advice, and in 1856 he set to work to make another and fuller abstract of the mass of data he had gathered.

And then a strange thing happened. After Darwin had been at work on his "abstract" about two years, but before he had published a line of it, there came to him one day a paper in manuscript, sent for his approval by a naturalist friend named Alfred Russel Wallace, who had been for some time at work in the East India Archipelago. He read the paper, and, to his amazement, found that it contained an outline of the same theory of "natural selection" which he himself had originated and for twenty years had worked upon. Working independently, on opposite sides of the globe, Darwin and Wallace had hit upon the same explanation of the cause of transmutation of species. "Were Wallace's paper an abstract of my unpublished manuscript of 1844," said Darwin, "it could not better express my ideas."

Here was a dilemma. To publish this paper with no word from Darwin would give Wallace priority, and wrest from Darwin the credit of a discovery which he had made years before his codiscoverer entered the field. Yet, on the other hand, could Darwin honorably do otherwise than publish his friend's paper and himself remain silent? It was a complication well calculated to try a man's soul. Darwin's was equal to the test. Keenly alive to the delicacy of the position, he placed the whole matter before his friends Hooker and Lyell, and left the decision as to a course of action absolutely to them. Needless to say, these great men did the one thing which insured full justice to all concerned. They counselled a joint publication, to include on the one hand Wallace's paper, and on the other an abstract of Darwin's ideas, in the exact form in which it had been outlined by the author in a letter to Asa Gray in the previous year—an abstract which was in Gray's hands before Wallace's paper was in existence. This joint production, together with a full statement of the facts of the case, was presented to the Linnaean Society of London by Hooker and Lyell on the evening of July 1, 1858, this being, by an odd coincidence, the twenty-first anniversary of the day on which Darwin had opened his journal to collect facts bearing on the "species question." Not often before in the history of science has it happened that a great theory has been nurtured in its author's brain through infancy and adolescence to its full legal majority before being sent out into the world.

Thus the fuse that led to the great powder-mine had been lighted. The explosion itself came more than a year later, in November, 1859, when Darwin, after thirteen months of further effort, completed the outline of his theory, which was at first begun as an abstract for the Linnaean Society, but which grew to the size of an independent volume despite his efforts at condensation, and which was given that ever–to–be–famous title, The Origin of Species by Means of Natural Selection, or the Preservation of Favored Races in the Struggle for Life. And what an explosion it was! The joint paper of 1858 had made a momentary flare, causing the hearers, as Hooker said, to "speak of it with bated breath," but beyond that it made no sensation. What the result was when the Origin itself appeared no one of our generation need be told. The rumble and roar that it made in the intellectual world have not yet altogether ceased to echo after more than forty years of reverberation.

NEW CHAMPIONS

To the Origin of Species, then, and to its author, Charles Darwin, must always be ascribed chief credit for that vast revolution in the fundamental beliefs of our race which has come about since 1859, and which made the second half of the century memorable. But it must not be overlooked that no such sudden metamorphosis could have been effected had it not been for the aid of a few notable lieutenants, who rallied to the standards of the leader immediately after the publication of the Origin. Darwin had all along felt the utmost confidence in the ultimate triumph of his ideas. "Our posterity," he declared, in a letter to Hooker, "will marvel as much about the current belief [in special creation] as we do about fossil shells having been thought to be created as we now see them." But he fully realized that for the present success of his theory of transmutation the championship of a few leaders of science was all–essential. He felt that if he could make converts of Hooker and Lyell and of Thomas Henry Huxley at once, all would be well.

His success in this regard, as in others, exceeded his expectations. Hooker was an ardent disciple from reading the proof–sheets before the book was published; Lyell renounced his former beliefs and fell into line a few months later; while Huxley, so soon as he had mastered the central idea of natural selection, marvelled that so simple yet all–potent a thought had escaped him so long, and then rushed eagerly into the fray, wielding the keenest dialectic blade that was drawn during the entire controversy. Then, too, unexpected recruits were found in Sir John Lubbock and John Tyndall, who carried the war eagerly into their respective territories; while Herbert Spencer, who had advocated a doctrine of transmutation on philosophic grounds some years before Darwin published the key to the mystery—and who himself had barely escaped independent discovery of that key—lent his masterful influence to the cause. In America the famous botanist Asa

Gray, who had long been a correspondent of Darwin's but whose advocacy of the new theory had not been anticipated, became an ardent propagandist; while in Germany Ernst Heinrich Haeckel, the youthful but already noted zoologist, took up the fight with equal enthusiasm.

Against these few doughty champions—with here and there another of less general renown—was arrayed, at the outset, practically all Christendom. The interest of the question came home to every person of intelligence, whatever his calling, and the more deeply as it became more and more clear how far–reaching are the real bearings of the doctrine of natural selection. Soon it was seen that should the doctrine of the survival of the favored races through the struggle for existence win, there must come with it as radical a change in man's estimate of his own position as had come in the day when, through the efforts of Copernicus and Galileo, the world was dethroned from its supposed central position in the universe. The whole conservative majority of mankind recoiled from this necessity with horror. And this conservative majority included not laymen merely, but a vast preponderance of the leaders of science also.

With the open–minded minority, on the other hand, the theory of natural selection made its way by leaps and bounds. Its delightful simplicity—which at first sight made it seem neither new nor important—coupled with the marvellous comprehensiveness of its implications, gave it a hold on the imagination, and secured it a hearing where other theories of transmutation of species had been utterly scorned. Men who had found Lamarck's conception of change through voluntary effort ridiculous, and the vaporings of the Vestiges altogether despicable, men whose scientific cautions held them back from Spencer's deductive argument, took eager hold of that tangible, ever–present principle of natural selection, and were led on and on to its goal. Hour by hour the attitude of the thinking world towards this new principle changed; never before was so great a revolution wrought so suddenly.

Nor was this merely because "the times were ripe" or "men's minds prepared for evolution." Darwin himself bears witness that this was not altogether so. All through the years in which he brooded this theory he sounded his scientific friends, and could find among them not one who acknowledged a doctrine of transmutation. The reaction from the stand–point of Lamarck and Erasmus Darwin and Goethe had been complete, and when Charles Darwin avowed his own conviction he expected always to have it met with ridicule or contempt. In 1857 there was but one man speaking with any large degree of authority in the world who openly avowed a belief in transmutation of species—that man being Herbert Spencer. But the Origin of Species came, as Huxley has said, like a flash in the darkness, enabling the benighted voyager to see the way. The score of years during which its author had waited and worked had been years well spent. Darwin had become, as he himself says, a veritable Croesus, "overwhelmed with his riches in facts"—facts of zoology, of selective artificial breeding, of geographical distribution of animals, of embryology, of paleontology. He had massed his facts about his theory, condensed them and recondensed, until his volume of five hundred pages was an encyclopaedia in scope. During those long years of musing he had thought out almost every conceivable objection to his theory, and in his book every such objection was stated with fullest force and candor, together with such reply as the facts at command might dictate. It was the force of those twenty years of effort of a master–mind that made the sudden breach in the breaswtork{sic} of current thought.

Once this breach was effected the work of conquest went rapidly on. Day by day squads of the enemy capitulated and struck their arms. By the time another score of years had passed the doctrine of evolution had become the working hypothesis of the scientific world. The revolution had been effected.

And from amid the wreckage of opinion and belief stands forth the figure of Charles Darwin, calm, imperturbable, serene; scatheless to ridicule, contumely, abuse; unspoiled by ultimate success; unsullied alike by the strife and the victory—take him for all in all, for character, for intellect, for what he was and what he did, perhaps the most Socratic figure of the century. When, in 1882, he died, friend and foe alike conceded that one of the greatest sons of men had rested from his labors, and all the world felt it fitting that the remains of Charles Darwin should be entombed in Westminster Abbey close beside the honored grave of Isaac Newton. Nor were there many who would dispute the justice of Huxley's estimate of his accomplishment: "He found a great truth trodden under foot. Reviled by bigots, and ridiculed by all the world, he lived long enough to see it, chiefly by his own efforts, irrefragably established in science, inseparably incorporated with the common thoughts of men, and only hated and feared by those who would revile but dare not."

THE ORIGIN OF THE FITTEST

Wide as are the implications of the great truth which Darwin and his co-workers established, however, it leaves quite untouched the problem of the origin of those "favored variations" upon which it operates. That such variations are due to fixed and determinate causes no one understood better than Darwin; but in his original exposition of his doctrine he made no assumption as to what these causes are. He accepted the observed fact of variation—as constantly witnessed, for example, in the differences between parents and offspring—and went ahead from this assumption.

But as soon as the validity of the principle of natural selection came to be acknowledged speculators began to search for the explanation of those variations which, for purposes of argument, had been provisionally called "spontaneous." Herbert Spencer had all along dwelt on this phase of the subject, expounding the Lamarckian conceptions of the direct influence of the environment (an idea which had especially appealed to Buffon and to Geoffroy Saint-Hilaire), and of effort in response to environment and stimulus as modifying the individual organism, and thus supplying the basis for the operation of natural selection. Haeckel also became an advocate of this idea, and presently there arose a so-called school of neo-Lamarckians, which developed particular strength and prominence in America under the leadership of Professors A. Hyatt and E. D. Cope.

But just as the tide of opinion was turning strongly in this direction, an utterly unexpected obstacle appeared in the form of the theory of Professor August Weismann, put forward in 1883, which antagonized the Lamarckian conception (though not touching the Darwinian, of which Weismann is a firm upholder) by denying that individual variations, however acquired by the mature organism, are transmissible. The flurry which this denial created has not yet altogether subsided, but subsequent observations seem to show that it was quite disproportionate to the real merits of the case. Notwithstanding Professor Weismann's objections, the balance of evidence appears to favor the view that the Lamarckian factor of acquired variations stands as the complement of the Darwinian factor of natural selection in effecting the transmutation of species.

Even though this partial explanation of what Professor Cope calls the "origin of the fittest" be accepted, there still remains one great life problem which the doctrine of evolution does not touch. The origin of species, genera, orders, and classes of beings through endless transmutations is in a sense explained; but what of the first term of this long series? Whence came that primordial organism whose transmuted descendants make up the existing faunas and floras of the globe?

There was a time, soon after the doctrine of evolution gained a hearing, when the answer to that question seemed to some scientists of authority to have been given by experiment. Recurring to a former belief, and repeating some earlier experiments, the director of the Museum of Natural History at Rouen, M. F. A. Pouchet, reached the conclusion that organic beings are spontaneously generated about us constantly, in the familiar processes of putrefaction, which were known to be due to the agency of microscopic bacteria. But in 1862 Louis Pasteur proved that this seeming spontaneous generation is in reality due to the existence of germs in the air. Notwithstanding the conclusiveness of these experiments, the claims of Pouchet were revived in England ten years later by Professor Bastian; but then the experiments of John Tyndall, fully corroborating the results of Pasteur, gave a final quietus to the claim of "spontaneous generation" as hitherto formulated.

There for the moment the matter rests. But the end is not yet. Fauna and flora are here, and, thanks to Lamarck and Wallace and Darwin, their development, through the operation of those "secondary causes" which we call laws of nature, has been proximally explained. The lowest forms of life have been linked with the highest in unbroken chains of descent. Meantime, through the efforts of chemists and biologists, the gap between the inorganic and the organic worlds, which once seemed almost infinite, has been constantly narrowed. Already philosophy can throw a bridge across that gap. But inductive science, which builds its own bridges, has not yet spanned the chasm, small though it appear. Until it shall have done so, the bridge of organic evolution is not quite complete; yet even as it stands to–day it is perhaps the most stupendous scientific structure of the nineteenth century.

VII. EIGHTEENTH–CENTURY MEDICINE

THE SYSTEM OF BOERHAAVE

At least two pupils of William Harvey distinguished themselves in medicine, Giorgio Baglivi (1669–1707), who has been called the "Italian Sydenham," and Hermann Boerhaave (1668–1738). The work of Baglivi was hardly begun before his early death removed one of the most promising of the early eighteenth–century physicians. Like Boerhaave, he represents a type of skilled, practical clinitian rather than the abstract scientist. One of his contributions to medical literature is the first accurate description of typhoid, or, as he calls it, mesenteric fever.

If for nothing else, Boerhaave must always be remembered as the teacher of Von Haller, but in his own day he was the widest known and the most popular teacher in the medical world. He was the idol of his pupils at Leyden, who flocked to his lectures in such numbers that it became necessary to "tear down the walls of Leyden to accommodate them." His fame extended not only all over Europe but to Asia, North America, and even into South America. A letter sent him from China was addressed to "Boerhaave in Europe." His teachings represent the best medical knowledge of his day, a high standard of morality, and a keen appreciation of the value of observation; and it was through such teachings imparted to his pupils and advanced by them, rather than to any new discoveries, that his name is important in medical history. His arrangement and classification of the different branches of medicine are interesting as representing the attitude of the medical profession towards these various branches at that time.

"In the first place we consider Life; then Health, afterwards Diseases; and lastly their several Remedies.

"Health the first general branch of Physic in our Institutions is termed Physiology, or the Animal Oeconomy; demonstrating the several Parts of the human Body, with their Mechanism and Actions.

"The second branch of Physic is called Pathology, treating of Diseases, their Differences, Causes and Effects, or Symptoms; by which the human Body is known to vary from its healthy state.

"The third part of Physic is termed Semiotica, which shows the Signs distinguishing between sickness and Health, Diseases and their Causes in the human Body; it also imports the State and Degrees of Health and Diseases, and presages their future Events.

"The fourth general branch of Physic is termed Hygiene, or Prophylaxis.

"The fifth and last part of Physic is called Therapeutica; which instructs us in the Nature, Preparation and uses of the Materia Medica; and the methods of applying the same, in order to cure Diseases and restore lost Health."[1]

From this we may gather that his general view of medicine was not unlike that taken at the present time.

Boerhaave's doctrines were arranged into a "system" by Friedrich Hoffmann, of Halle (1660–1742), this system having the merit of being simple and more easily comprehended than many others. In this system forces were considered inherent in matter, being expressed as mechanical movements, and determined by mass, number, and weight. Similarly, forces express themselves in the body by movement, contraction, and relaxation, etc., and life itself is movement, "particularly movement of the heart." Life and death are, therefore, mechanical phenomena, health is determined by regularly recurring movements, and disease by irregularity of them. The body is simply a large hydraulic machine, controlled by "the aether" or "sensitive soul," and the chief centre of this soul lies in the medulla.

In the practical application of medicines to diseases Hoffman used simple remedies, frequently with happy results, for whatever the medical man's theory may be he seldom has the temerity to follow it out logically, and use the remedies indicated by his theory to the exclusion of long–established, although perhaps purely empirical, remedies. Consequently, many vague theorists have been excellent practitioners, and Hoffman was one of these. Some of the remedies he introduced are still in use, notably the spirits of ether, or "Hoffman's anodyne."

ANIMISTS, VITALISTS, AND ORGANICISTS

Besides Hoffman's system of medicine, there were numerous others during the eighteenth century, most of which are of no importance whatever; but three, at least, that came into existence and disappeared during the century are worthy of fuller notice. One of these, the Animists, had for its chief exponent Georg Ernst Stahl of "phlogiston" fame; another, the Vitalists, was championed by Paul Joseph Barthez (1734–1806); and the third was the Organicists. This last, while agreeing with the other two that vital activity cannot be explained by the laws of physics and chemistry, differed in not believing that life "was due to some spiritual entity," but rather to the structure of the body itself.

The Animists taught that the soul performed functions of ordinary life in man, while the life of lower animals was controlled by ordinary mechanical principles. Stahl supported this theory ardently, sometimes violently, at times declaring that there were "no longer any doctors, only mechanics and chemists." He denied that chemistry had anything to do with medicine, and, in the main, discarded anatomy as useless to the medical man. The soul, he thought, was the source of all vital movement; and the immediate cause of death was not disease but the direct action of the soul. When through some lesion, or because the machinery of the body has become unworkable, as in old age, the soul leaves the body and death is produced. The soul ordinarily selects the channels of the circulation, and the contractile parts, as the route for influencing the body. Hence in fever the pulse is quickened, due to the increased activity of the soul, and convulsions and spasmodic movements in disease are due, to the, same cause. Stagnation of the, blood was supposed to be a fertile cause of diseases, and such diseases were supposed to arise mostly from "plethora"—an all–important element in Stahl's therapeutics. By many this theory is regarded as an attempt on the part of the pious Stahl to reconcile medicine and theology in a way satisfactory to both physicians and theologians, but, like many conciliatory attempts, it was violently opposed by both doctors and ministers.

A belief in such a theory would lead naturally to simplicity in therapeutics, and in this respect at least Stahl was consistent. Since the soul knew more about the body than any physician could know, Stahl conceived that it would be a hinderance rather than a help for the physician to interfere with complicated doses of medicine. As he advanced in age this view of the administration of drugs grew upon him, until after rejecting quinine, and finally opium, he at last used only salt and water in treating his patients. From this last we may judge that his "system," if not doing much good, was at least doing little harm.

The theory of the Vitalists was closely allied to that of the Animists, and its most important representative, Paul Joseph Barthez, was a cultured and eager scientist. After an eventful and varied career as physician, soldier, editor, lawyer, and philosopher in turn, he finally returned to the field of medicine, was made consulting physician by Napoleon in 1802, and died in Paris four years later.

The theory that he championed was based on the assumption that there was a "vital principle," the nature of which was unknown, but which differed from the thinking mind, and was the cause of the phenomena of life. This "vital principle" differed from the soul, and was not exhibited in human beings alone, but even in animals and plants. This force, or whatever it might be called, was supposed to be present everywhere in the body, and all diseases were the results of it.

The theory of the Organicists, like that of the Animists and Vitalists, agreed with the other two that vital activity could not be explained by the laws of physics and chemistry, but, unlike them, it held that it was a part of the structure of the body itself. Naturally the practical physicians were more attracted by this tangible doctrine than by vague theories "which converted diseases into unknown derangements of some equally unknown 'principle.' "

It is perhaps straining a point to include this brief description of these three schools of medicine in the history of the progress of the science. But, on the whole, they were negatively at least prominent factors in directing true progress along its proper channel, showing what courses were not to be pursued. Some one has said that science usually stumbles into the right course only after stumbling into all the wrong ones; and if this be only partially true, the wrong ones still play a prominent if not a very creditable part. Thus the medical systems of William Cullen (1710–1790), and John Brown (1735–1788), while doing little towards the actual advancement of scientific medicine, played so conspicuous a part in so wide a field that the "Brunonian system" at least must be given some little attention.

According to Brown's theory, life, diseases, and methods of cure are explained by the property of "excitability." All exciting powers were supposed to be stimulating, the apparent debilitating effects of some being due to a deficiency in the amount of stimulus. Thus "the whole phenomena of life, health, as well as disease, were supposed to consist of stimulus and nothing else." This theory created a great stir in the medical world, and partisans and opponents sprang up everywhere. In Italy it was enthusiastically supported; in England it was strongly opposed; while in Scotland riots took place between the opposing factions. Just why this system should have created any stir, either for or against it, is not now apparent.

Like so many of the other "theorists" of his century, Brown's practical conclusions deduced from his theory (or perhaps in spite of it) were generally beneficial to medicine, and some of them extremely valuable in the treatment of diseases. He first advocated the modern stimulant, or "feeding treatment" of fevers, and first recognized the usefulness of animal soups and beef–tea in certain diseases.

THE SYSTEM OF HAHNEMANN

Just at the close of the century there came into prominence the school of homoeopathy, which was destined to influence the practice of medicine very materially and to outlive all the other eighteenth–century schools. It was founded by Christian Samuel Friedrich Hahnemann (1755–1843), a most remarkable man, who, after propounding a theory in his younger days which was at least as reasonable as most of the existing theories, had the misfortune to outlive his usefulness and lay his doctrine open to ridicule by the unreasonable teachings of his dotage,

Hahnemann rejected all the teachings of morbid anatomy and pathology as useless in practice, and propounded his famous "similia similibus curantur"—that all diseases were to be cured by medicine which in health produced symptoms dynamically similar to the disease under treatment. If a certain medicine produced a headache when given to a healthy person, then this medicine was indicated in case of headaches, etc. At the present time such a theory seems crude enough, but in the latter part of the eighteenth century almost any theory was as good as the ones propounded by Animists, Vitalists, and other such schools. It certainly had the very commendable feature of introducing simplicity in the use of drugs in place of the complicated prescriptions then in vogue. Had Hahnemann stopped at this point he could not have been held up to the indefensible ridicule that was brought upon him, with considerable justice, by his later theories. But he lived onto propound his extraordinary theory of "potentiality"—that medicines gained strength by being diluted—and his even more extraordinary theory that all chronic diseases are caused either by the itch, syphilis, or fig–wart disease, or are brought on by medicines.

At the time that his theory of potentialities was promulgated, the medical world had gone mad in its administration of huge doses of compound mixtures of drugs, and any reaction against this was surely an improvement. In short, no medicine at all was much better than the heaping doses used in common practice; and hence one advantage, at least, of Hahnemann's methods. Stated briefly, his theory was that if a tincture be reduced to one–fiftieth in strength, and this again reduced to one–fiftieth, and this process repeated up to thirty such dilutions, the potency of such a medicine will be increased by each dilution, Hahnemann himself preferring the weakest, or, as he would call it, the strongest dilution. The absurdity of such a theory is apparent when it is understood that long before any drug has been raised to its thirtieth dilution it has been so reduced in quantity that it cannot be weighed, measured, or recognized as being present in the solution at all by any means known to chemists. It is but just to modern followers of homoeopathy to say that while most of them advocate small dosage, they do not necessarily follow the teachings of Hahnemann in this respect, believing that the theory of the dose "has nothing more to do with the original law of cure than the psora (itch) theory has; and that it was one of the later creations of Hahnemann's mind."

Hahnemann's theory that all chronic diseases are derived from either itch, syphilis, or fig—wart disease is no longer advocated by his followers, because it is so easily disproved, particularly in the case of itch. Hahnemann taught that fully three—quarters of all diseases were caused by "itch struck in," and yet it had been demonstrated long before his day, and can be demonstrated any time, that itch is simply a local skin disease caused by a small parasite.

JENNER AND VACCINATION

All advances in science have a bearing, near or remote, on the welfare of our race; but it remains to credit to the closing decade of the eighteenth century a discovery which, in its power of direct and immediate benefit to humanity, surpasses any other discovery of this or any previous epoch. Needless to say, I refer to Jenner's discovery of the method of preventing smallpox by inoculation with the virus of cow—pox. It detracts nothing from the merit of this discovery to say that the preventive power of accidental inoculation had long been rumored among the peasantry of England. Such vague, unavailing half—knowledge is often the forerunner of fruitful discovery.

To all intents and purposes Jenner's discovery was original and unique. Nor, considered as a perfect method, was it in any sense an accident. It was a triumph of experimental science. The discoverer was no novice in scientific investigation, but a trained observer, who had served a long apprenticeship in scientific observation under no less a scientist than the celebrated John Hunter. At the age of twenty—one Jenner had gone to London to pursue his medical studies, and soon after he proved himself so worthy a pupil that for two years he remained a member of Hunter's household as his favorite pupil. His taste for science and natural history soon attracted the attention of Sir Joseph Banks, who intrusted him with the preparation of the zoological specimens brought back by Captain Cook's expedition in 1771. He performed this task so well that he was offered the position of naturalist to the second expedition, but declined it, preferring to take up the practice of his profession in his native town of Berkeley.

His many accomplishments and genial personality soon made him a favorite both as a physician and in society. He was a good singer, a fair violinist and flute–player, and a very successful writer of prose and verse. But with all his professional and social duties he still kept up his scientific investigations, among other things making some careful observations on the hibernation of hedgehogs at the instigation of Hunter, the results of which were laid before the Royal Society. He also made quite extensive investigations as to the geological formations and fossils found in his neighborhood.

Even during his student days with Hunter he had been much interested in the belief, current in the rural districts of Gloucestershire, of the antagonism between cow–pox and small–pox, a person having suffered from cow–pox being immuned to small–pox. At various times Jenner had mentioned the subject to Hunter, and he was constantly making inquiries of his fellow–practitioners as to their observations and opinions on the subject. Hunter was too fully engrossed in other pursuits to give the matter much serious attention, however, and Jenner's brothers of the profession gave scant credence to the rumors, although such rumors were common enough.

At this time the practice of inoculation for preventing small–pox, or rather averting the severer forms of the disease, was widely practised. It was customary, when there was a mild case of the disease, to take some of the virus from the patient and inoculate persons who had never had the disease, producing a similar attack in them. Unfortunately there were many objections to this practice. The inoculated patient frequently developed a virulent form of the disease and died; or if he recovered, even after a mild attack, he was likely to be "pitted" and disfigured. But, perhaps worst of all, a patient so inoculated became the source of infection to others, and it sometimes happened that disastrous epidemics were thus brought about. The case was a most perplexing one, for the awful scourge of small–pox hung perpetually over the head of every person who had not already suffered and recovered from it. The practice of inoculation was introduced into England by Lady Mary Wortley Montague (1690–1762), who had seen it practised in the East, and who announced her intention of "introducing it into England in spite of the doctors."

From the fact that certain persons, usually milkmaids, who had suffered from cow–pox seemed to be immuned to small–pox, it would seem a very simple process of deduction to discover that cow–pox inoculation was the solution of the problem of preventing the disease. But there was another form of disease which, while closely resembling cow–pox and quite generally confounded with it, did not produce immunity. The confusion of these two forms of the disease had constantly misled investigations as to the possibility of either of them immunizing against smallpox, and the confusion of these two diseases for a time led Jenner to question the possibility of doing so. After careful investigations, however, he reached the conclusion that there was a difference in the effects of the two diseases, only one of which produced immunity from small-pox.

"There is a disease to which the horse, from his state of domestication, is frequently subject," wrote Jenner, in his famous paper on vaccination. "The farriers call it the grease. It is an inflammation and swelling in the heel, accompanied at its commencement with small cracks or fissures, from which issues a limpid fluid possessing properties of a very peculiar kind. This fluid seems capable of generating a disease in the human body (after it has undergone the modification I shall presently speak of) which bears so strong a resemblance to small-pox that I think it highly probable it may be the source of that disease.

"In this dairy country a great number of cows are kept, and the office of milking is performed indiscriminately by men and maid servants. One of the former having been appointed to apply dressings to the heels of a horse affected with the malady I have mentioned, and not paying due attention to cleanliness, incautiously bears his part in milking the cows with some particles of the infectious matter adhering to his fingers. When this is the case it frequently happens that a disease is communicated to the cows, and from the cows to the dairy-maids, which spreads through the farm until most of the cattle and domestics feel its unpleasant consequences. This disease has obtained the name of Cow-Pox. It appears on the nipples of the cows in the form of irregular pustules. At their first appearance they are commonly of a palish blue, or rather of a color somewhat approaching to livid, and are surrounded by an inflammation. These pustules, unless a timely remedy be applied, frequently degenerate into phagedenic ulcers, which prove extremely troublesome. The animals become indisposed, and the secretion of milk is much lessened. Inflamed spots now begin to appear on different parts of the hands of the domestics employed in milking, and sometimes on the wrists, which run on to suppuration, first assuming the appearance of the small vesications produced by a burn. Most commonly they appear about the joints of the fingers and at their extremities; but whatever parts are affected, if the situation will admit the superficial suppurations put on a circular form with their edges more elevated than their centre and of a color distinctly approaching to blue. Absorption takes place, and tumors appear in each axilla. The system becomes affected, the pulse is quickened; shiverings, succeeded by heat, general lassitude, and pains about the loins and limbs, with vomiting, come on. The head is painful, and the patient is now and then even affected with delirium. These symptoms, varying in their degrees of violence, generally continue from one day to three or four, leaving ulcerated sores about the hands which, from the sensibility of the parts, are very troublesome and commonly heal slowly, frequently becoming phagedenic, like those from which they sprang. During the progress of the disease the lips, nostrils, eyelids, and other parts of the body are sometimes affected with sores; but these evidently arise from their being heedlessly rubbed or scratched by the patient's infected fingers. No eruptions on the skin have followed the decline of the feverish symptoms in any instance that has come under my inspection, one only excepted, and in this case a very few appeared on the arms: they were very minute, of a vivid red color, and soon died away without advancing to maturation, so that I cannot determine whether they had any connection with the preceding symptoms.

"Thus the disease makes its progress from the horse (as I conceive) to the nipple of the cow, and from the cow to the human subject.

"Morbid matter of various kinds, when absorbed into the system, may produce effects in some degree similar; but what renders the cow-pox virus so extremely singular is that the person that has been thus affected is forever after secure from the infection of small-pox, neither exposure to the variolous effluvia nor the insertion of the matter into the skin producing this distemper."[2]

In 1796 Jenner made his first inoculation with cowpox matter, and two months later the same subject was inoculated with small-pox matter. But, as Jenner had predicted, no attack of small-pox followed. Although fully convinced by this experiment that the case was conclusively proven, he continued his investigations, waiting two years before publishing his discovery. Then, fortified by indisputable proofs, he gave it to the world. The immediate effects of his announcement have probably never been equalled in the history of scientific discovery, unless, perhaps, in the single instance of the discovery of anaesthesia. In Geneva and Holland clergymen advocated the practice of vaccination from their pulpits; in some of the Latin countries religious processions were formed for receiving vaccination; Jenner's birthday was celebrated as a feast in Germany; and the first child vaccinated in Russia was named "Vaccinov" and educated at public expense. In six years the discovery had penetrated to the most remote corners of civilization; it had even reached some savage nations. And in a few years small—pox had fallen from the position of the most dreaded of all diseases to that of being practically the only disease for which a sure and easy preventive was known.

Honors were showered upon Jenner from the Old and the New World, and even Napoleon, the bitter hater of the English, was among the others who honored his name. On one occasion Jenner applied to the Emperor for the release of certain Englishmen detained in France. The petition was about to be rejected when the name of the petitioner was mentioned. "Ah," said Napoleon, "we can refuse nothing to that name!"

It is difficult for us of to—day clearly to conceive the greatness of Jenner's triumph, for we can only vaguely realize what a ruthless and ever—present scourge smallpox had been to all previous generations of men since history began. Despite all efforts to check it by medication and by direct inoculation, it swept now and then over the earth as an all—devastating pestilence, and year by year it claimed one—tenth of all the beings in Christendom by death as its average quota of victims. "From small—pox and love but few remain free," ran the old saw. A pitted face was almost as much a matter of course a hundred years ago as a smooth one is to—day.

Little wonder, then, that the world gave eager acceptance to Jenner's discovery. No urging was needed to induce the majority to give it trial; passengers on a burning ship do not hold aloof from the life—boats. Rich and poor, high and low, sought succor in vaccination and blessed the name of their deliverer. Of all the great names that were before the world in the closing days of the century, there was perhaps no other one at once so widely known and so uniformly reverenced as that of the great English physician Edward Jenner. Surely there was no other one that should be recalled with greater gratitude by posterity.

VIII. NINETEENTH—CENTURY MEDICINE

PHYSICAL DIAGNOSIS

Although Napoleon Bonaparte, First Consul, was not lacking in self—appreciation, he probably did not realize that in selecting a physician for his own needs he was markedly influencing the progress of medical science as a whole. Yet so strangely are cause and effect adjusted in human affairs that this simple act of the First Consul had that very unexpected effect. For the man chosen was the envoy of a new method in medical practice, and the fame which came to him through being physician to the First Consul, and subsequently to the Emperor, enabled him to promulgate the method in a way otherwise impracticable. Hence the indirect but telling value to medical science of Napoleon's selection.

The physician in question was Jean Nicolas de Corvisart. His novel method was nothing more startling than the now−familiar procedure of tapping the chest of a patient to elicit sounds indicative of diseased tissues within. Every one has seen this done commonly enough in our day, but at the beginning of the century Corvisart, and perhaps some of his pupils, were probably the only physicians in the world who resorted to this simple and useful procedure. Hence Napoleon's surprise when, on calling in Corvisart, after becoming somewhat dissatisfied with his other physicians Pinel and Portal, his physical condition was interrogated in this strange manner. With characteristic shrewdness Bonaparte saw the utility of the method, and the physician who thus attempted to substitute scientific method for guess−work in the diagnosis of disease at once found favor in his eyes and was installed as his regular medical adviser.

For fifteen years before this Corvisart had practised percussion, as the chest−tapping method is called, without succeeding in convincing the profession of its value. The method itself, it should be added, had not originated with Corvisart, nor did the French physician for a moment claim it as his own. The true originator of the practice was the German physician Avenbrugger, who published a book about it as early as 1761. This book had even been translated into French, then the language of international communication everywhere, by Roziere de la Chassagne, of Montpellier, in 1770; but no one other than Corvisart appears to have paid any attention to either original or translation. It was far otherwise, however, when Corvisart translated Avenbrugger's work anew, with important additions of his own, in 1808.

"I know very well how little reputation is allotted to translator and commentators," writes Corvisart, "and I might easily have elevated myself to the rank of an author if I had elaborated anew the doctrine of Avenbrugger and published an independent work on percussion. In this way, however, I should have sacrificed the name of Avenbrugger to my own vanity, a thing which I am unwilling to do. It is he, and the beautiful invention which of right belongs to him, that I desire to recall to life."[1]

120

By this time a reaction had set in against the metaphysical methods in medicine that had previously been so alluring; the scientific spirit of the time was making itself felt in medical practice; and this, combined with Corvisart's fame, brought the method of percussion into immediate and well–deserved popularity. Thus was laid the foundation for the method of so–called physical diagnosis, which is one of the corner–stones of modern medicine.

The method of physical diagnosis as practised in our day was by no means completed, however, with the work of Corvisart. Percussion alone tells much less than half the story that may be elicited from the organs of the chest by proper interrogation. The remainder of the story can only be learned by applying the ear itself to the chest, directly or indirectly. Simple as this seems, no one thought of practising it for some years after Corvisart had shown the value of percussion.

Then, in 1815, another Paris physician, Rene Theophile Hyacinthe Laennec, discovered, almost by accident, that the sound of the heart–beat could be heard surprisingly through a cylinder of paper held to the ear and against the patient's chest. Acting on the hint thus received, Laennec substituted a hollow cylinder of wood for the paper, and found himself provided with an instrument through which not merely heart sounds but murmurs of the lungs in respiration could be heard with almost startling distinctness.

The possibility of associating the varying chest sounds with diseased conditions of the organs within appealed to the fertile mind of Laennec as opening new vistas in therapeutics, which he determined to enter to the fullest extent practicable. His connection with the hospitals of Paris gave him full opportunity in this direction, and his labors of the next few years served not merely to establish the value of the new method as an aid to diagnosis, but laid the foundation also for the science of morbid anatomy. In 1819 Laennec published the results of his labors in a work called Traite d'Auscultation Mediate,[2] a work which forms one of the landmarks of scientific medicine. By mediate auscultation is meant, of course, the interrogation of the chest with the aid of the little instrument already referred to, an instrument which its originator thought hardly worth naming until various barbarous appellations were applied to it by others, after which Laennec decided to call it the stethoscope, a name which it has ever since retained.

In subsequent years the form of the stethoscope, as usually employed, was modified and its value augmented by a binauricular attachment, and in very recent years a further improvement has been made through application of the principle of the telephone; but the essentials of auscultation with the stethoscope were established in much detail by Laennec, and the honor must always be his of thus taking one of the longest single steps by which practical medicine has in our century acquired the right to be considered a rational science. Laennec's efforts cost him his life, for he died in 1826 of a lung disease acquired in the course of his hospital practice; but even before this his fame was universal, and the value of his method had been recognized all over the world. Not long after, in 1828, yet another French physician, Piorry, perfected the method of percussion by introducing the custom of tapping, not the chest directly, but the finger or a small metal or hard-rubber plate held against the chest-mediate percussion, in short. This perfected the methods of physical diagnosis of diseases of the chest in all essentials; and from that day till this percussion and auscultation have held an unquestioned place in the regular armamentarium of the physician.

Coupled with the new method of physical diagnosis in the effort to substitute knowledge for guess-work came the studies of the experimental physiologists—in particular, Marshall Hall in England and Francois Magendie in France; and the joint efforts of these various workers led presently to the abandonment of those severe and often irrational depletive methods—blood-letting and the like—that had previously dominated medical practice. To this end also the "statistical method," introduced by Louis and his followers, largely contributed; and by the close of the first third of our century the idea was gaining ground that the province of therapeutics is to aid nature in combating disease, and that this may often be accomplished better by simple means than by the heroic measures hitherto thought necessary. In a word, scientific empiricism was beginning to gain a hearing in medicine as against the metaphysical preconceptions of the earlier generations.

PARASITIC DISEASES

I have just adverted to the fact that Napoleon Bonaparte, as First Consul and as Emperor, was the victim of a malady which caused him to seek the advice of the most distinguished physicians of Paris. It is a little shocking to modern sensibilities to read that these physicians, except Corvisart, diagnosed the distinguished patient's malady as "gale repercutee"—that is to say, in idiomatic English, the itch "struck in." It is hardly necessary to say that no physician of today would make so inconsiderate a diagnosis in the case of a royal patient. If by any chance a distinguished patient were afflicted with the itch, the sagacious physician would carefully hide the fact behind circumlocutions and proceed to eradicate the disease with all despatch. That the physicians of Napoleon did otherwise is evidence that at the beginning of the century the disease in question enjoyed a very different status. At that time itch, instead of being a most plebeian malady, was, so to say, a court disease. It enjoyed a circulation, in high circles and in low, that modern therapeutics has quite denied it; and the physicians of the time gave it a fictitious added importance by ascribing to its influence the existence of almost any obscure malady that came under their observation. Long after Napoleon's time gale continued to hold this proud distinction. For example, the imaginative Dr. Hahnemann did not hesitate to affirm, as a positive maxim, that three—fourths of all the ills that flesh is heir to were in reality nothing but various forms of "gale repercutee."

All of which goes to show how easy it may be for a masked pretender to impose on credulous humanity, for nothing is more clearly established in modern knowledge than the fact that "gale repercutee" was simply a name to hide a profound ignorance; no such disease exists or ever did exist. Gale itself is a sufficiently tangible reality, to be sure, but it is a purely local disease of the skin, due to a perfectly definite cause, and the dire internal conditions formerly ascribed to it have really no causal connection with it whatever. This definite cause, as every one nowadays knows, is nothing more or less than a microscopic insect which has found lodgment on the skin, and has burrowed and made itself at home there. Kill that insect and the disease is no more; hence it has come to be an axiom with the modern physician that the itch is one of the three or four diseases that he positively is able to cure, and that very speedily. But it was far otherwise with the physicians of the first third of our century, because to them the cause of the disease was an absolute mystery.

It is true that here and there a physician had claimed to find an insect lodged in the skin of a sufferer from itch, and two or three times the claim had been made that this was the cause of the malady, but such views were quite ignored by the general profession, and in 1833 it was stated in an authoritative medical treatise that the "cause of gale is absolutely unknown." But even at this time, as it curiously happened, there were certain ignorant laymen who had attained to a bit of medical knowledge that was withheld from the inner circles of the profession. As the peasantry of England before Jenner had known of the curative value of cow—pox over small—pox, so the peasant women of Poland had learned that the annoying skin disease from which they suffered was caused by an almost invisible insect, and, furthermore, had acquired the trick of dislodging the pestiferous little creature with the point of a needle. From them a youth of the country, F. Renucci by name, learned the open secret. He conveyed it to Paris when he went there to study medicine, and in 1834 demonstrated it to his master Alibert. This physician, at first sceptical, soon was convinced, and gave out the discovery to the medical world with an authority that led to early acceptance.

Now the importance of all this, in the present connection, is not at all that it gave the clew to the method of cure of a single disease. What makes the discovery epochal is the fact that it dropped a brand—new idea into the medical ranks—an idea destined, in the long—run, to prove itself a veritable bomb—the idea, namely, that a minute and quite unsuspected animal parasite may be the cause of a well—known, widely prevalent, and important human disease. Of course the full force of this idea could only be appreciated in the light of later knowledge; but even at the time of its coming it sufficed to give a great impetus to that new medical knowledge, based on microscopical studies, which had but recently been made accessible by the inventions of the lens—makers. The new knowledge clarified one very turbid medical pool and pointed the way to the clarification of many others.

Almost at the same time that the Polish medical student was demonstrating the itch mite in Paris, it chanced, curiously enough, that another medical student, this time an Englishman, made an analogous discovery of perhaps even greater importance. Indeed, this English discovery in its initial stages slightly antedated the other, for it was in 1833 that the student in question, James Paget, interne in St. Bartholomew's Hospital, London, while dissecting the muscular tissues of a human subject, found little specks of extraneous matter, which, when taken to the professor of comparative anatomy, Richard Owen, were ascertained, with the aid of the microscope, to be the cocoon of a minute and hitherto unknown insect. Owen named the insect Trichina spiralis. After the discovery was published it transpired that similar specks had been observed by several earlier investigators, but no one had previously suspected or, at any rate, demonstrated their nature. Nor was the full story of the trichina made out for a long time after Owen's discovery. It was not till 1847 that the American anatomist Dr. Joseph Leidy found the cysts of trichina in the tissues of pork; and another decade or so elapsed after that before German workers, chief among whom were Leuckart, Virchow, and Zenker, proved that the parasite gets into the human system through ingestion of infected pork, and that it causes a definite set of symptoms of disease which hitherto had been mistaken for rheumatism, typhoid fever, and other maladies. Then the medical world was agog for a time over the subject of trichinosis; government inspection of pork was established in some parts of Germany; American pork was excluded altogether from France; and the whole subject thus came prominently to public attention. But important as the trichina parasite proved on its own account in the end, its greatest importance, after all, was in the share it played in directing attention at the time of its discovery in 1833 to the subject of microscopic parasites in general.

The decade that followed that discovery was a time of great activity in the study of microscopic organisms and microscopic tissues, and such men as Ehrenberg and Henle and Bory Saint–Vincent and Kolliker and Rokitansky and Remak and Dujardin were widening the bounds of knowledge of this new subject with details that cannot be more than referred to here. But the crowning achievement of the period in this direction was the discovery made by the German, J. L. Schoenlein, in 1839, that a very common and most distressing disease of the scalp, known as favus, is really due to the presence and growth on the scalp of a vegetable organism of microscopic size. Thus it was made clear that not merely animal but also vegetable organisms of obscure, microscopic species have causal relations to the diseases with which mankind is afflicted. This knowledge of the parasites was another long step in the direction of scientific medical knowledge; but the heights to which this knowledge led were not to be scaled, or even recognized, until another generation of workers had entered the field.

PAINLESS SURGERY

Meantime, in quite another field of medicine, events were developing which led presently to a revelation of greater immediate importance to humanity than any other discovery that had come in the century, perhaps in any field of science whatever. This was the discovery of the pain–dispelling power of the vapor of sulphuric ether inhaled by a patient undergoing a surgical operation. This discovery came solely out of America, and it stands curiously isolated, since apparently no minds in any other country were trending towards it even vaguely. Davy, in England, had indeed originated the method of medication by inhalation, and earned out some most interesting experiments fifty years earlier, and it was doubtless his experiments with nitrous oxide gas that gave the clew to one of the American investigators; but this was the sole contribution of preceding generations to the subject, and since the beginning of the century, when Davy turned his attention to other matters, no one had made the slightest advance along the same line until an American dentist renewed the investigation.

In view of the sequel, Davy's experiments merit full attention. Here is his own account of them, as written in 1799:

"Immediately after a journey of one hundred and twenty–six miles, in which I had no sleep the preceding night, being much exhausted, I respired seven quarts of nitrous oxide gas for near three minutes. It produced the usual pleasurable effects and slight muscular motion. I continued exhilarated for some minutes afterwards, but in half an hour found myself neither more nor less exhausted than before the experiment. I had a great propensity to sleep.

"To ascertain with certainty whether the more extensive action of nitrous oxide compatible with life was capable of producing debility, I resolved to breathe the gas for such a time, and in such quantities, as to produce excitement equal in duration and superior in intensity to that occasioned by high intoxication from opium or alcohol.

"To habituate myself to the excitement, and to carry it on gradually, on December 26th I was enclosed in an air–tight breathing–box, of the capacity of about nine and one–half cubic feet, in the presence of Dr. Kinglake. After I had taken a situation in which I could by means of a curved thermometer inserted under the arm, and a stop–watch, ascertain the alterations in my pulse and animal heat, twenty quarts of nitrous oxide were thrown into the box.

"For three minutes I experienced no alteration in my sensations, though immediately after the introduction of the nitrous oxide the smell and taste of it were very evident. In four minutes I began to feel a slight glow in the cheeks and a generally diffused warmth over the chest, though the temperature of the box was not quite 50 degrees. . . . In twenty–five minutes the animal heat was 100 degrees, pulse 124. In thirty minutes twenty quarts more of gas were introduced.

"My sensations were now pleasant; I had a generally diffused warmth without the slightest moisture of the skin, a sense of exhilaration similar to that produced by a small dose of wine, and a disposition to muscular motion and to merriment.

"In three–quarters of an hour the pulse was 104 and the animal heat not 99.5 degrees, the temperature of the chamber 64 degrees. The pleasurable feelings continued to increase, the pulse became fuller and slower, till in about an hour it was 88, when the animal heat was 99 degrees. Twenty quarts more of air were admitted. I had now a great disposition to laugh, luminous points seemed frequently to pass before my eyes, my hearing was certainly more acute, and I felt a pleasant lightness and power of exertion in my muscles. In a short time the symptoms became stationary; breathing was rather oppressed, and on account of the great desire for action rest was painful.

"I now came out of the box, having been in precisely an hour and a quarter. The moment after I began to respire twenty quarts of unmingled nitrous oxide. A thrilling extending from the chest to the extremities was almost immediately produced. I felt a sense of tangible extension highly pleasurable in every limb; my visible impressions were dazzling and apparently magnified, I heard distinctly every sound in the room, and was perfectly aware of my situation. By degrees, as the pleasurable sensations increased, I lost all connection with external things; trains of vivid visible images rapidly passed through my mind and were connected with words in such a manner as to produce perceptions perfectly novel.

"I existed in a world of newly connected and newly modified ideas. I theorized; I imagined that I made discoveries. When I was awakened from this semi–delirious trance by Dr. Kinglake, who took the bag from my mouth, indignation and pride were the first feelings produced by the sight of persons about me. My emotions were enthusiastic and sublime; and for a minute I walked about the room perfectly regardless of what was said to me. As I recovered my former state of mind, I felt an inclination to communicate the discoveries I had made during the experiment. I endeavored to recall the ideas—they were feeble and indistinct; one collection of terms, however, presented itself, and, with most intense belief and prophetic manner, I exclaimed to Dr. Kinglake, 'Nothing exists but thoughts!—the universe is composed of impressions, ideas, pleasures, and pains.' "[3]

From this account we see that Davy has anaesthetized himself to a point where consciousness of surroundings was lost, but not past the stage of exhilaration. Had Dr. Kinglake allowed the inhaling–bag to remain in Davy's mouth for a few moments longer complete insensibility would have followed. As it was, Davy appears to have realized that sensibility was dulled, for he adds this illuminative suggestion: "As nitrous oxide in its extensive operation appears capable of destroying physical pain, it may probably be used with advantage during surgical operations in which no great effusion of blood takes place."[4]

Unfortunately no one took advantage of this suggestion at the time, and Davy himself became interested in other fields of science and never returned to his physiological studies, thus barely missing one of the greatest discoveries in the entire field of science. In the generation that followed no one seems to have thought of putting Davy's suggestion to the test, and the surgeons of Europe had acknowledged with one accord that all hope of finding a means to render operations painless must be utterly abandoned—that the surgeon's knife must ever remain a synonym for slow and indescribable torture. By an odd coincidence it chanced that Sir Benjamin Brodie, the acknowledged leader of English surgeons, had publicly expressed this as his deliberate though regretted opinion at a time when the quest which he considered futile had already led to the most brilliant success in America, and while the announcement of the discovery, which then had no transatlantic cable to convey it, was actually on its way to the Old World.

The American dentist just referred to, who was, with one exception to be noted presently, the first man in the world to conceive that the administration of a definite drug might render a surgical operation painless and to give the belief application was Dr. Horace Wells, of Hartford, Connecticut. The drug with which he experimented was nitrous oxide—the same that Davy had used; the operation that he rendered painless was no more important than the extraction of a tooth—yet it sufficed to mark a principle; the year of the experiment was 1844.

The experiments of Dr. Wells, however, though important, were not sufficiently demonstrative to bring the matter prominently to the attention of the medical world. The drug with which he experimented proved not always reliable, and he himself seems ultimately to have given the matter up, or at least to have relaxed his efforts. But meantime a friend, to whom he had communicated his belief and expectations, took the matter up, and with unremitting zeal carried forward experiments that were destined to lead to more tangible results. This friend was another dentist, Dr. W. T. G. Morton, of Boston, then a young man full of youthful energy and enthusiasm. He seems to have felt that the drug with which Wells had experimented was not the most practicable one for the purpose, and so for several months he experimented with other allied drugs, until finally he hit upon sulphuric ether, and with this was able to make experiments upon animals, and then upon patients in the dental chair, that seemed to him absolutely demonstrative.

Full of eager enthusiasm, and absolutely confident of his results, he at once went to Dr. J. C. Warren, one of the foremost surgeons of Boston, and asked permission to test his discovery decisively on one of the patients at the Boston Hospital during a severe operation. The request was granted; the test was made on October 16, 1846, in the presence of several of the foremost surgeons of the city and of a body of medical students. The patient slept quietly while the surgeon's knife was plied, and awoke to astonished comprehension that the ordeal was over. The impossible, the miraculous, had been accomplished.[5]

Swiftly as steam could carry it—slowly enough we should think it to–day—the news was heralded to all the world. It was received in Europe with incredulity, which vanished before repeated experiments. Surgeons were loath to believe that ether, a drug that had long held a place in the subordinate armamentarium of the physician, could accomplish such a miracle. But scepticism vanished before the tests which any surgeon might make, and which surgeons all over the world did make within the next few weeks. Then there came a lingering outcry from a few surgeons, notably some of the Parisians, that the shock of pain was beneficial to the patient, hence that anaesthesia—as Dr. Oliver Wendell Holmes had christened the new method—was a procedure not to be advised. Then, too, there came a hue–and–cry from many a pulpit that pain was God–given, and hence, on moral grounds, to be clung to rather than renounced. But the outcry of the antediluvians of both hospital and pulpit quickly received its quietus; for soon it was clear that the patient who did not suffer the shock of pain during an operation rallied better than the one who did so suffer, while all humanity outside the pulpit cried shame to the spirit that would doom mankind to suffer needless agony. And so within a few months after that initial operation at the Boston Hospital in 1846, ether had made good its conquest of pain throughout the civilized world. Only by the most active use of the imagination can we of this present day realize the full meaning of that victory.

It remains to be added that in the subsequent bickerings over the discovery—such bickerings as follow every great advance—two other names came into prominent notice as sharers in the glory of the new method. Both these were Americans—the one, Dr. Charles T. Jackson, of Boston; the other, Dr. Crawford W. Long, of Alabama. As to Dr. Jackson, it is sufficient to say that he seems to have had some vague inkling of the peculiar properties of ether before Morton's discovery. He even suggested the use of this drug to Morton, not knowing that Morton had already tried it; but this is the full measure of his association with the discovery. Hence it is clear that Jackson's claim to equal share with Morton in the discovery was unwarranted, not to say absurd.

Dr. Long's association with the matter was far different and altogether honorable. By one of those coincidences so common in the history of discovery, he was experimenting with ether as a pain–destroyer simultaneously with Morton, though neither so much as knew of the existence of the other. While a medical student he had once inhaled ether for the intoxicant effects, as other medical students were wont to do, and when partially under influence of the drug he had noticed that a chance blow to his shins was painless. This gave him the idea that ether might be used in surgical operations; and in subsequent years, in the course of his practice in a small Georgia town, he put the idea into successful execution. There appears to be no doubt whatever that he performed successful minor operations under ether some two or three years before Morton's final demonstration; hence that the merit of first using the drug, or indeed any drug, in this way belongs to him. But, unfortunately, Dr. Long did not quite trust the evidence of his own experiments. Just at that time the medical journals were full of accounts of experiments in which painless operations were said to be performed through practice of hypnotism, and Dr. Long feared that his own success might be due to an incidental hypnotic influence rather than to the drug. Hence he delayed announcing his apparent discovery until he should have opportunity for further tests—and opportunities did not come every day to the country practitioner. And while he waited, Morton anticipated him, and the discovery was made known to the world without his aid. It was a true scientific caution that actuated Dr. Long to this delay, but the caution cost him the credit, which might otherwise have been his, of giving to the world one of the greatest blessings—dare we not, perhaps, say the very greatest?—that science has ever conferred upon humanity.

A few months after the use of ether became general, the Scotch surgeon Sir J. Y. Simpson[6] discovered that another drug, chloroform, could be administered with similar effects; that it would, indeed, in many cases produce anaesthesia more advantageously even than ether. From that day till this surgeons have been more or less divided in opinion as to the relative merits of the two drugs; but this fact, of course, has no bearing whatever upon the merit of the first discovery of the method of anaesthesia. Even had some other drug subsequently quite banished ether, the honor of the discovery of the beneficent method of anaesthesia would have been in no wise invalidated. And despite all cavillings, it is unequivocally established that the man who gave that method to the world was William T. G. Morton.

PASTEUR AND THE GERM THEORY OF DISEASE

The discovery of the anaesthetic power of drugs was destined presently, in addition to its direct beneficences, to aid greatly in the progress of scientific medicine, by facilitating those experimental studies of animals from which, before the day of anaesthesia, many humane physicians were withheld, and which in recent years have led to discoveries of such inestimable value to humanity. But for the moment this possibility was quite overshadowed by the direct benefits of anaesthesia, and the long strides that were taken in scientific medicine during the first fifteen years after Morton's discovery were mainly independent of such aid. These steps were taken, indeed, in a field that at first glance might seem to have a very slight connection with medicine. Moreover, the chief worker in the field was not himself a physician. He was a chemist, and the work in which he was now engaged was the study of alcoholic fermentation in vinous liquors. Yet these studies paved the way for the most important advances that medicine has made in any century towards the plane of true science; and to this man more than to any other single individual—it might almost be said more than to all other individuals—was due this wonderful advance. It is almost superfluous to add that the name of this marvellous chemist was Louis Pasteur.

The studies of fermentation which Pasteur entered upon in 1854 were aimed at the solution of a controversy that had been waging in the scientific world with varying degrees of activity for a quarter of a century. Back in the thirties, in the day of the early enthusiasm over the perfected microscope, there had arisen a new interest in the minute forms of life which Leeuwenhoek and some of the other early workers with the lens had first described, and which now were shown to be of almost universal prevalence. These minute organisms had been studied more or less by a host of observers, but in particular by the Frenchman Cagniard Latour and the German of cell–theory fame, Theodor Schwann. These men, working independently, had reached the conclusion, about 1837, that the micro–organisms play a vastly more important role in the economy of nature than any one previously had supposed. They held, for example, that the minute specks which largely make up the substance of yeast are living vegetable organisms, and that the growth of these organisms is the cause of the important and familiar process of fermentation. They even came to hold, at least tentatively, the opinion that the somewhat similar micro–organisms to be found in all putrefying matter, animal or vegetable, had a causal relation to the process of putrefaction.

This view, particularly as to the nature of putrefaction, was expressed even more outspokenly a little later by the French botanist Turpin. Views so supported naturally gained a following; it was equally natural that so radical an innovation should be antagonized. In this case it chanced that one of the most dominating scientific minds of the time, that of Liebig, took a firm and aggressive stand against the new doctrine. In 1839 he promulgated his famous doctrine of fermentation, in which he stood out firmly against any "vitalistic" explanation of the phenomena, alleging that the presence of micro—organisms in fermenting and putrefying substances was merely incidental, and in no sense causal. This opinion of the great German chemist was in a measure substantiated by experiments of his compatriot Helmholtz, whose earlier experiments confirmed, but later ones contradicted, the observations of Schwann, and this combined authority gave the vitalistic conception a blow from which it had not rallied at the time when Pasteur entered the field. Indeed, it was currently regarded as settled that the early students of the subject had vastly over—estimated the importance of micro—organisms.

And so it came as a new revelation to the generality of scientists of the time, when, in 1857 and the succeeding half—decade, Pasteur published the results of his researches, in which the question had been put to a series of altogether new tests, and brought to unequivocal demonstration.

He proved that the micro—organisms do all that his most imaginative predecessors had suspected, and more. Without them, he proved, there would be no fermentation, no putrefaction—no decay of any tissues, except by the slow process of oxidation. It is the microscopic yeast—plant which, by seizing on certain atoms of the molecule, liberates the remaining atoms in the form of carbonic—acid and alcohol, thus effecting fermentation; it is another microscopic plant—a bacterium, as Devaine had christened it—which in a similar way effects the destruction of organic molecules, producing the condition which we call putrefaction. Pasteur showed, to the amazement of biologists, that there are certain forms of these bacteria which secure the oxygen which all organic life requires, not from the air, but by breaking up unstable molecules in which oxygen is combined; that putrefaction, in short, has its foundation in the activities of these so—called anaerobic bacteria.

In a word, Pasteur showed that all the many familiar processes of the decay of organic tissues are, in effect, forms of fermentation, and would not take place at all except for the presence of the living micro—organisms. A piece of meat, for example, suspended in an atmosphere free from germs, will dry up gradually, without the slightest sign of putrefaction, regardless of the temperature or other conditions to which it may have been subjected. Let us witness one or two series of these experiments as presented by Pasteur himself in one of his numerous papers before the Academy of Sciences.

EXPERIMENTS WITH GRAPE SUGAR

"In the course of the discussion which took place before the Academy upon the subject of the generation of ferments properly so–called, there was a good deal said about that of wine, the oldest fermentation known. On this account I decided to disprove the theory of M. Fremy by a decisive experiment bearing solely upon the juice of grapes.

"I prepared forty flasks of a capacity of from two hundred and fifty to three hundred cubic centimetres and filled them half full with filtered grape–must, perfectly clear, and which, as is the case of all acidulated liquids that have been boiled for a few seconds, remains uncontaminated although the curved neck of the flask containing them remain constantly open during several months or years.

"In a small quantity of water I washed a part of a bunch of grapes, the grapes and the stalks together, and the stalks separately. This washing was easily done by means of a small badger's–hair brush. The washing–water collected the dust upon the surface of the grapes and the stalks, and it was easily shown under the microscope that this water held in suspension a multitude of minute organisms closely resembling either fungoid spores, or those of alcoholic Yeast, or those of Mycoderma vini, etc. This being done, ten of the forty flasks were preserved for reference; in ten of the remainder, through the straight tube attached to each, some drops of the washing–water were introduced; in a third series of ten flasks a few drops of the same liquid were placed after it had been boiled; and, finally, in the ten remaining flasks were placed some drops of grape–juice taken from the inside of a perfect fruit. In order to carry out this experiment, the straight tube of each flask was drawn out into a fine and firm point in the lamp, and then curved. This fine and closed point was filed round near the end and inserted into the grape while resting upon some hard substance. When the point was felt to touch the support of the grape it was by a slight pressure broken off at the point file mark. Then, if care had been taken to create a slight vacuum in the flask, a drop of the juice of the grape got into it, the filed point was withdrawn, and the aperture immediately closed in the alcohol lamp. This decreased pressure of the atmosphere in the flask was obtained by the following means: After warming the sides of the flask either in the hands or in the lamp–flame, thus causing a small quantity of air to be driven out of the end of the curved neck, this end was closed in the lamp. After the flask was cooled, there was a tendency to suck in the drop of grape–juice in the manner just described.

"The drop of grape–juice which enters into the flask by this suction ordinarily remains in the curved part of the tube, so that to mix it with the must it was necessary to incline the flask so as to bring the must into contact with the juice and then replace the flask in its normal position. The four series of comparative experiments produced the following results:

"The first ten flasks containing the grape–must boiled in pure air did not show the production of any organism. The grape–must could possibly remain in them for an indefinite number of years. Those in the second series, containing the water in which the grapes had been washed separately and together, showed without exception an alcoholic fermentation which in several cases began to appear at the end of forty–eight hours when the experiment took place at ordinary summer temperature. At the same time that the yeast appeared, in the form of white traces, which little by little united themselves in the form of a deposit on the sides of all the flasks, there were seen to form little flakes of Mycellium, often as a single fungoid growth or in combination, these fungoid growths being quite independent of the must or of any alcoholic yeast. Often, also, the Mycoderma vini appeared after some days upon the surface of the liquid. The Vibria and the lactic ferments properly so called did not appear on account of the nature of the liquid.

"The third series of flasks, the washing–water in which had been previously boiled, remained unchanged, as in the first series. Those of the fourth series, in which was the juice of the interior of the grapes, remained equally free from change, although I was not always able, on account of the delicacy of the experiment, to eliminate every chance of error. These experiments cannot leave the least doubt in the mind as to the following facts:

Grape–must, after heating, never ferments on contact with the air, when the air has been deprived of the germs which it ordinarily holds in a state of suspension.

"The boiled grape–must ferments when there is introduced into it a very small quantity of water in which the surface of the grapes or their stalks have been washed.

"The grape—must does not ferment when this washing—water has been boiled and afterwards cooled.

"The grape—must does not ferment when there is added to it a small quantity of the juice of the inside of the grape.

"The yeast, therefore, which causes the fermentation of the grapes in the vintage—tub comes from the outside and not from the inside of the grapes. Thus is destroyed the hypothesis of MM. Trecol and Fremy, who surmised that the albuminous matter transformed itself into yeast on account of the vital germs which were natural to it. With greater reason, therefore, there is no longer any question of the theory of Liebig of the transformation of albuminoid matter into ferments on account of the oxidation."

FOREIGN ORGANISMS AND THE WORT OF BEER

"The method which I have just followed," Pasteur continues, "in order to show that there exists a correlation between the diseases of beer and certain microscopic organisms leaves no room for doubt, it seems to me, in regard to the principles I am expounding.

"Every time that the microscope reveals in the leaven, and especially in the active yeast, the production of organisms foreign to the alcoholic yeast properly so called, the flavor of the beer leaves something to be desired, much or little, according to the abundance and the character of these little germs. Moreover, when a finished beer of good quality loses after a time its agreeable flavor and becomes sour, it can be easily shown that the alcoholic yeast deposited in the bottles or the casks, although originally pure, at least in appearance, is found to be contaminated gradually with these filiform or other ferments. All this can be deduced from the facts already given, but some critics may perhaps declare that these foreign ferments are the consequences of the diseased condition, itself produced by unknown causes.

"Although this gratuitous hypothesis may be difficult to uphold, I will endeavor to corroborate the preceding observations by a clearer method of investigation. This consists in showing that the beer never has any unpleasant taste in all cases when the alcoholic ferment properly so called is not mixed with foreign ferments; that it is the same in the case of wort, and that wort, liable to changes as it is, can be preserved unaltered if it is kept from those microscopic parasites which find in it a suitable nourishment and a field for growth.

"The employment of this second method has, moreover, the advantage of proving with certainty the proposition that I advanced at first—namely, that the germs of these organisms are derived from the dust of the atmosphere, carried about and deposited upon all objects, or scattered over the utensils and the materials used in a brewery—materials naturally charged with microscopic germs, and which the various operations in the store–rooms and the malt–house may multiply indefinitely.

"Let us take a glass flask with a long neck of from two hundred and fifty to three hundred cubic centimetres capacity, and place in it some wort, with or without hops, and then in the flame of a lamp draw out the neck of the flask to a fine point, afterwards heating the liquid until the steam comes out of the end of the neck. It can then be allowed to cool without any other precautions; but for additional safety there can be introduced into the little point a small wad of asbestos at the moment that the flame is withdrawn from beneath the flask. Before thus placing the asbestos it also can be passed through the flame, as well as after it has been put into the end of the tube. The air which then first re–enters the flask will thus come into contact with the heated glass and the heated liquid, so as to destroy the vitality of any dust germs that may exist in the air. The air itself will re–enter very gradually, and slowly enough to enable any dust to be taken up by the drop of water which the air forces up the curvature of the tube. Ultimately the tube will be dry, but the re–entering of the air will be so slow that the particles of dust will fall upon the sides of the tube. The experiments show that with this kind of vessel, allowing free communication with the air, and the dust not being allowed to enter, the dust will not enter at all events for a period of ten or twelve years, which has been the longest period devoted to these trials; and the liquid, if it were naturally limpid, will not be in the least polluted neither on its surface nor in its mass, although the outside of the flask may become thickly coated with dust. This is a most irrefutable proof of the impossibility of dust getting inside the flask.

"The wort thus prepared remains uncontaminated indefinitely, in spite of its susceptibility to change when exposed to the air under conditions which allow it to gather the dusty particles which float in the atmosphere. It is the same in the case of urine, beef–tea, and grape–must, and generally with all those putrefactable and fermentable liquids which have the property when heated to boiling–point of destroying the vitality of dust germs."[7]

There was nothing in these studies bearing directly upon the question of animal diseases, yet before they were finished they had stimulated progress in more than one field of pathology. At the very outset they sufficed to start afresh the inquiry as to the role played by micro–organisms in disease. In particular they led the French physician Devaine to return to some interrupted studies which he had made ten years before in reference to the animal disease called anthrax, or splenic fever, a disease that cost the farmers of Europe millions of francs annually through loss of sheep and cattle. In 1850 Devaine had seen multitudes of bacteria in the blood of animals who had died of anthrax, but he did not at that time think of them as having a causal relation to the disease. Now, however, in 1863, stimulated by Pasteur's new revelations regarding the power of bacteria, he returned to the subject, and soon became convinced, through experiments by means of inoculation, that the microscopic organisms he had discovered were the veritable and the sole cause of the infectious disease anthrax.

The publication of this belief in 1863 aroused a furor of controversy. That a microscopic vegetable could cause a virulent systemic disease was an idea altogether too startling to be accepted in a day, and the generality of biologists and physicians demanded more convincing proofs than Devaine as yet was able to offer.

Naturally a host of other investigators all over the world entered the field. Foremost among these was the German Dr. Robert Koch, who soon corroborated all that Devaine had observed, and carried the experiments further in the direction of the cultivation of successive generations of the bacteria in artificial media, inoculations being made from such pure cultures of the eighth generation, with the astonishing result that animals thus inoculated succumbed to the disease.

Such experiments seem demonstrative, yet the world was unconvinced, and in 1876, while the controversy was still at its height, Pasteur was prevailed upon to take the matter in hand. The great chemist was becoming more and more exclusively a biologist as the years passed, and in recent years his famous studies of the silk–worm diseases, which he proved due to bacterial infection, and of the question of spontaneous generation, had given him unequalled resources in microscopical technique. And so when, with the aid of his laboratory associates Duclaux and Chamberland and Roux, he took up the mooted anthrax question the scientific world awaited the issue with bated breath. And when, in 1877, Pasteur was ready to report on his studies of anthrax, he came forward with such a wealth of demonstrative experiments—experiments the rigid accuracy of which no one would for a moment think of questioning—going to prove the bacterial origin of anthrax, that scepticism was at last quieted for all time to come.

Henceforth no one could doubt that the contagious disease anthrax is due exclusively to the introduction into an animal's system of a specific germ—a microscopic plant—which develops there. And no logical mind could have a reasonable doubt that what is proved true of one infectious disease would some day be proved true also of other, perhaps of all, forms of infectious maladies.

Hitherto the cause of contagion, by which certain maladies spread from individual to individual, had been a total mystery, quite unillumined by the vague terms "miasm," "humor," "virus," and the like cloaks of ignorance. Here and there a prophet of science, as Schwann and Henle, had guessed the secret; but guessing, in science, is far enough from knowing. Now, for the first time, the world KNEW, and medicine had taken another gigantic stride towards the heights of exact science.

LISTER AND ANTISEPTIC SURGERY

Meantime, in a different though allied field of medicine there had been a complementary growth that led to immediate results of even more practical importance. I mean the theory and practice of antisepsis in surgery. This advance, like the other, came as a direct outgrowth of Pasteur's fermentation studies of alcoholic beverages, though not at the hands of Pasteur himself. Struck by the boundless implications of Pasteur's revelations regarding the bacteria, Dr. Joseph Lister (the present Lord Lister), then of Glasgow, set about as early as 1860 to make a wonderful application of these ideas. If putrefaction is always due to bacterial development, he argued, this must apply as well to living as to dead tissues; hence the putrefactive changes which occur in wounds and after operations on the human subject, from which blood–poisoning so often follows, might be absolutely prevented if the injured surfaces could be kept free from access of the germs of decay.

In the hope of accomplishing this result, Lister began experimenting with drugs that might kill the bacteria without injury to the patient, and with means to prevent further access of germs once a wound was freed from them. How well he succeeded all the world knows; how bitterly he was antagonized for about a score of years, most of the world has already forgotten. As early as 1867 Lister was able to publish results pointing towards success in his great project; yet so incredulous were surgeons in general that even some years later the leading surgeons on the Continent had not so much as heard of his efforts. In 1870 the soldiers of Paris died, as of old, of hospital gangrene; and when, in 1871, the French surgeon Alphonse Guerin, stimulated by Pasteur's studies, conceived the idea of dressing wounds with cotton in the hope of keeping germs from entering them, he was quite unaware that a British contemporary had preceded him by a full decade in this effort at prevention and had made long strides towards complete success. Lister's priority, however, and the superiority of his method, were freely admitted by the French Academy of Sciences, which in 1881 officially crowned his achievement, as the Royal Society of London had done the year before.

By this time, to be sure, as everybody knows, Lister's new methods had made their way everywhere, revolutionizing the practice of surgery and practically banishing from the earth maladies that hitherto had been the terror of the surgeon and the opprobrium of his art. And these bedside studies, conducted in the end by thousands of men who had no knowledge of microscopy, had a large share in establishing the general belief in the causal relation that micro–organisms bear to disease, which by about the year 1880 had taken possession of the medical world. But they did more; they brought into equal prominence the idea that, the cause of a diseased condition being known, it maybe possible as never before to grapple with and eradicate that condition.

PREVENTIVE INOCULATION

The controversy over spontaneous generation, which, thanks to Pasteur and Tyndall, had just been brought to a termination, made it clear that no bacterium need be feared where an antecedent bacterium had not found lodgment; Listerism in surgery had now shown how much might be accomplished towards preventing the access of germs to abraded surfaces of the body and destroying those that already had found lodgment there. As yet, however, there was no inkling of a way in which a corresponding onslaught might be made upon those other germs which find their way into the animal organism by way of the mouth and the nostrils, and which, as was now clear, are the cause of those contagious diseases which, first and last, claim so large a proportion of mankind for their victims. How such means might be found now became the anxious thought of every imaginative physician, of every working microbiologist.

As it happened, the world was not kept long in suspense. Almost before the proposition had taken shape in the minds of the other leaders, Pasteur had found a solution. Guided by the empirical success of Jenner, he, like many others, had long practised inoculation experiments, and on February 9, 1880, he announced to the French Academy of Sciences that he had found a method of so reducing the virulence of a disease germ that when introduced into the system of a susceptible animal it produced only a mild form of the disease, which, however, sufficed to protect against the usual virulent form exactly as vaccinia protects against small-pox. The particular disease experimented with was that infectious malady of poultry known familiarly as "chicken cholera." In October of the same year Pasteur announced the method by which this "attenuation of the virus," as he termed it, had been brought about—by cultivation of the disease germs in artificial media, exposed to the air, and he did not hesitate to assert his belief that the method would prove "susceptible of generalization"—that is to say, of application to other diseases than the particular one in question.

Within a few months he made good this prophecy, for in February, 1881, he announced to the Academy that with the aid, as before, of his associates MM. Chamberland and Roux, he had produced an attenuated virus of the anthrax microbe by the use of which, as he affirmed with great confidence, he could protect sheep, and presumably cattle, against that fatal malady. "In some recent publications," said Pasteur, "I announced the first case of the attenuation of a virus by experimental methods only. Formed of a special microbe of an extreme minuteness, this virus may be multiplied by artificial culture outside the animal body. These cultures, left alone without any possible external contamination, undergo, in the course of time, modifications of their virulency to a greater or less extent. The oxygen of the atmosphere is said to be the chief cause of these attenuations—that is, this lessening of the facilities of multiplication of the microbe; for it is evident that the difference of virulence is in some way associated with differences of development in the parasitic economy.

"There is no need to insist upon the interesting character of these results and the deductions to be made therefrom. To seek to lessen the virulence by rational means would be to establish, upon an experimental basis, the hope of preparing from an active virus, easily cultivated either in the human or animal body, a vaccine-virus of restrained development capable of preventing the fatal effects of the former. Therefore, we have applied all our energies to investigate the possible generalizing action of atmospheric oxygen in the attenuation of virus.

"The anthrax virus, being one that has been most carefully studied, seemed to be the first that should attract our attention. Every time, however, we encountered a difficulty. Between the microbe of chicken cholera and the microbe of anthrax there exists an essential difference which does not allow the new experiment to be verified by the old. The microbes of chicken cholera do not, in effect, seem to resolve themselves, in their culture, into veritable germs. The latter are merely cells, or articulations always ready to multiply by division, except when the particular conditions in which they become true germs are known.

"The yeast of beer is a striking example of these cellular productions, being able to multiply themselves indefinitely without the apparition of their original spores. There exist many mucedines (Mucedinae?) of tubular mushrooms, which in certain conditions of culture produce a chain of more or less spherical cells called Conidae. The latter, detached from their branches, are able to reproduce themselves in the form of cells, without the appearance, at least with a change in the conditions of culture, of the spores of their respective mucedines. These vegetable organisms can be compared to plants which are cultivated by slipping, and to produce which it is not necessary to have the fruits or the seeds of the mother plant.

The anthrax bacterium, in its artificial cultivation, behaves very differently. Its mycelian filaments, if one may so describe them, have been produced scarcely for twenty–four or forty–eight hours when they are seen to transform themselves, those especially which are in free contact with the air, into very refringent corpuscles, capable of gradually isolating themselves into true germs of slight organization. Moreover, observation shows that these germs, formed so quickly in the culture, do not undergo, after exposure for a time to atmospheric air, any change either in their vitality or their virulence. I was able to present to the Academy a tube containing some spores of anthrax bacteria produced four years ago, on March 21, 1887. Each year the germination of these little corpuscles has been tried, and each year the germination has been accomplished with the same facility and the same rapidity as at first. Each year also the virulence of the new cultures has been tested, and they have not shown any visible falling off. Therefore, how can we experiment with the action of the air upon the anthrax virus with any expectation of making it less virulent?

"The crucial difficulty lies perhaps entirely in this rapid reproduction of the bacteria germs which we have just related. In its form of a filament, and in its multiplication by

141

division, is not this organism at all points comparable with the microbe of the chicken cholera?

"That a germ, properly so called, that a seed, does not suffer any modification on account of the air is easily conceived; but it is conceivable not less easily that if there should be any change it would occur by preference in the case of a mycelian fragment. It is thus that a slip which may have been abandoned in the soil in contact with the air does not take long to lose all vitality, while under similar conditions a seed is preserved in readiness to reproduce the plant. If these views have any foundation, we are led to think that in order to prove the action of the air upon the anthrax bacteria it will be indispensable to submit to this action the mycelian development of the minute organism under conditions where there cannot be the least admixture of corpuscular germs. Hence the problem of submitting the bacteria to the action of oxygen comes back to the question of presenting entirely the formation of spores. The question being put in this way, we are beginning to recognize that it is capable of being solved.

"We can, in fact, prevent the appearance of spores in the artificial cultures of the anthrax parasite by various artifices. At the lowest temperature at which this parasite can be cultivated—that is to say, about +16 degrees Centigrade—the bacterium does not produce germs—at any rate, for a very long time. The shapes of the minute microbe at this lowest limit of its development are irregular, in the form of balls and pears—in a word, they are monstrosities—but they are without spores. In the last regard also it is the same at the highest temperatures at which the parasite can be cultivated, temperatures which vary slightly according to the means employed. In neutral chicken bouillon the bacteria cannot be cultivated above 45 degrees. Culture, however, is easy and abundant at 42 to 43 degrees, but equally without any formation of spores. Consequently a culture of mycelian bacteria can be kept entirely free from germs while in contact with the open air at a temperature of from 42 to 43 degrees Centigrade. Now appear the three remarkable results. After about one month of waiting the culture dies—that is to say, if put into a fresh bouillon it becomes absolutely sterile.

"So much for the life and nutrition of this organism. In respect to its virulence, it is an extraordinary fact that it disappears entirely after eight days' culture at 42 to 43 degrees Centigrade, or, at any rate, the cultures are innocuous for the guinea–pig, the rabbit, and the sheep, the three kinds of animals most apt to contract anthrax. We are thus able to obtain, not only the attenuation of the virulence, but also its complete suppression by a simple method of cultivation. Moreover, we see also the possibility of preserving and cultivating the terrible microbe in an inoffensive state. What is it that happens in these eight days at 43 degrees that suffices to take away the virulence of the bacteria? Let us remember that the microbe of chicken cholera dies in contact with the air, in a period somewhat protracted, it is true, but after successive attenuations. Are we justified in thinking that it ought to be the same in regard to the microbe of anthrax? This hypothesis is confirmed by experiment. Before the disappearance of its virulence the anthrax microbe passes through various degrees of attenuation, and, moreover, as is also the case with the microbe of chicken cholera, each of these attenuated states of virulence can be obtained by cultivation. Moreover, since, according to one of our recent Communications, anthrax is not recurrent, each of our attenuated anthrax microbes is, for the better–developed microbe, a vaccine—that is to say, a virus producing a less–malignant malady. What, therefore, is easier than to find in these a virus that will infect with anthrax sheep, cows, and horses, without killing them, and ultimately capable of warding off the mortal malady? We have practised this experiment with great success upon sheep, and when the season comes for the assembling of the flocks at Beauce we shall try the experiment on a larger scale.

"Already M. Toussaint has announced that sheep can be saved by preventive inoculations; but when this able observer shall have published his results; on the subject of which we have made such exhaustive studies, as yet unpublished, we shall be able to see the whole difference which exists between the two methods—the uncertainty of the one and the certainty of the other. That which we announce has, moreover, the very great advantage of resting upon the existence of a poison vaccine cultivable at will, and which can be increased indefinitely in the space of a few hours without having recourse to infected blood."[8]

This announcement was immediately challenged in a way that brought it to the attention of the entire world. The president of an agricultural society, realizing the enormous importance of the subject, proposed to Pasteur that his alleged discovery should be submitted to a decisive public test. He proposed to furnish a drove of fifty sheep half of which were to be inoculated with the attenuated virus of Pasteur. Subsequently all the sheep were to be inoculated with virulent virus, all being kept together in one pen under precisely the same conditions. The "protected" sheep were to remain healthy; the unprotected ones to die of anthrax; so read the terms of the proposition. Pasteur accepted the challenge; he even permitted a change in the programme by which two goats were substituted for two of the sheep, and ten cattle added, stipulating, however, that since his experiments had not yet been extended to cattle these should not be regarded as falling rigidly within the terms of the test.

It was a test to try the soul of any man, for all the world looked on askance, prepared to deride the maker of so preposterous a claim as soon as his claim should be proved baseless. Not even the fame of Pasteur could make the public at large, lay or scientific, believe in the possibility of what he proposed to accomplish. There was time for all the world to be informed of the procedure, for the first "preventive" inoculation—or vaccination, as Pasteur termed it—was made on May 5th, the second on May 17th, and another interval of two weeks must elapse before the final inoculations with the unattenuated virus. Twenty–four sheep, one goat, and five cattle were submitted to the preliminary vaccinations. Then, on May 31 st, all sixty of the animals were inoculated, a protected and unprotected one alternately, with an extremely virulent culture of anthrax microbes that had been in Pasteur's laboratory since 1877. This accomplished, the animals were left together in one enclosure to await the issue.

Two days later, June 2d, at the appointed hour of rendezvous, a vast crowd, composed of veterinary surgeons, newspaper correspondents, and farmers from far and near, gathered to witness the closing scenes of this scientific tourney. What they saw was one of the most dramatic scenes in the history of peaceful science—a scene which, as Pasteur declared afterwards, "amazed the assembly." Scattered about the enclosure, dead, dying, or manifestly sick unto death, lay the unprotected animals, one and all, while each and every "protected" animal stalked unconcernedly about with every appearance of perfect health. Twenty of the sheep and the one goat were already dead; two other sheep expired under the eyes of the spectators; the remaining victims lingered but a few hours longer. Thus in a manner theatrical enough, not to say tragic, was proclaimed the unequivocal victory of science. Naturally enough, the unbelievers struck their colors and surrendered without terms; the principle of protective vaccination, with a virus experimentally prepared in the laboratory, was established beyond the reach of controversy.

That memorable scientific battle marked the beginning of a new era in medicine. It was a foregone conclusion that the principle thus established would be still further generalized; that it would be applied to human maladies; that in all probability it would grapple successfully, sooner or later, with many infectious diseases. That expectation has advanced rapidly towards realization. Pasteur himself made the application to the human subject in the disease hydrophobia in 1885, since which time that hitherto most fatal of maladies has largely lost its terrors. Thousands of persons bitten by mad dogs have been snatched from the fatal consequences of that mishap by this method at the Pasteur Institute in Paris, and at the similar institutes, built on the model of this parent one, that have been established all over the world in regions as widely separated as New York and Nha–Trang.

SERUM–THERAPY

In the production of the rabies vaccine Pasteur and his associates developed a method of attenuation of a virus quite different from that which had been employed in the case of the vaccines of chicken cholera and of anthrax. The rabies virus was inoculated into the system of guinea–pigs or rabbits and, in effect, cultivated in the systems of these animals. The spinal cord of these infected animals was found to be rich in the virus, which rapidly became attenuated when the cord was dried in the air. The preventive virus, of varying strengths, was made by maceration of these cords at varying stages of desiccation. This cultivation of a virus within the animal organism suggested, no doubt, by the familiar Jennerian method of securing small–pox vaccine, was at the same time a step in the direction of a new therapeutic procedure which was destined presently to become of all–absorbing importance—the method, namely, of so–called serum–therapy, or the treatment of a disease with the blood serum of an animal that has been subjected to protective inoculation against that disease.

The possibility of such a method was suggested by the familiar observation, made by Pasteur and numerous other workers, that animals of different species differ widely in their susceptibility to various maladies, and that the virus of a given disease may become more and more virulent when passed through the systems of successive individuals of one species, and, contrariwise, less and less virulent when passed through the systems of successive individuals of another species. These facts suggested the theory that the blood of resistant animals might contain something directly antagonistic to the virus, and the hope that this something might be transferred with curative effect to the blood of an infected susceptible animal. Numerous experimenters all over the world made investigations along the line of this alluring possibility, the leaders perhaps being Drs. Behring and Kitasato, closely followed by Dr. Roux and his associates of the Pasteur Institute of Paris. Definite results were announced by Behring in 1892 regarding two important diseases—tetanus and diphtheria—but the method did not come into general notice until 1894, when Dr. Roux read an epoch-making paper on the subject at the Congress of Hygiene at Buda-Pesth.

In this paper Dr. Roux, after adverting to the labors of Behring, Ehrlich, Boer, Kossel, and Wasserman, described in detail the methods that had been developed at the Pasteur Institute for the development of the curative serum, to which Behring had given the since-familiar name antitoxine. The method consists, first, of the cultivation, for some months, of the diphtheria bacillus (called the Klebs-Loeffler bacillus, in honor of its discoverers) in an artificial bouillon, for the development of a powerful toxine capable of giving the disease in a virulent form.

This toxine, after certain details of mechanical treatment, is injected in small but increasing doses into the system of an animal, care being taken to graduate the amount so that the animal does not succumb to the disease. After a certain course of this treatment it is found that a portion of blood serum of the animal so treated will act in a curative way if injected into the blood of another animal, or a human patient, suffering with diphtheria. In other words, according to theory, an antitoxine has been developed in the system of the animal subjected to the progressive inoculations of the diphtheria toxine. In Dr. Roux's experience the animal best suited for the purpose is the horse, though almost any of the domesticated animals will serve the purpose.

But Dr. Roux's paper did not stop with the description of laboratory methods. It told also of the practical application of the serum to the treatment of numerous cases of diphtheria in the hospitals of Paris—applications that had met with a gratifying measure of success. He made it clear that a means had been found of coping successfully with what had been one of the most virulent and intractable of the diseases of childhood. Hence it was not strange that his paper made a sensation in all circles, medical and lay alike.

Physicians from all over the world flocked to Paris to learn the details of the open secret, and within a few months the new serum-therapy had an acknowledged standing with the medical profession everywhere. What it had accomplished was regarded as but an earnest of what the new method might accomplish presently when applied to the other infectious diseases.

Efforts at such applications were immediately begun in numberless directions—had, indeed, been under way in many a laboratory for some years before. It is too early yet to speak of the results in detail. But enough has been done to show that this method also is susceptible of the widest generalization. It is not easy at the present stage to sift that which is tentative from that which will be permanent; but so great an authority as Behring does not hesitate to affirm that today we possess, in addition to the diphtheria antitoxine, equally specific antitoxines of tetanus, cholera, typhus fever, pneumonia, and tuberculosis—a set of diseases which in the aggregate account for a startling proportion of the general death–rate. Then it is known that Dr. Yersin, with the collaboration of his former colleagues of the Pasteur Institute, has developed, and has used with success, an antitoxine from the microbe of the plague which recently ravaged China.

Dr. Calmette, another graduate of the Pasteur Institute, has extended the range of the serum–therapy to include the prevention and treatment of poisoning by venoms, and has developed an antitoxine that has already given immunity from the lethal effects of snake bites to thousands of persons in India and Australia.

Just how much of present promise is tentative, just what are the limits of the methods—these are questions for the future to decide. But, in any event, there seems little question that the serum treatment will stand as the culminating achievement in therapeutics of our century. It is the logical outgrowth of those experimental studies with the microscope begun by our predecessors of the thirties, and it represents the present culmination of the rigidly experimental method which has brought medicine from a level of fanciful empiricism to the plane of a rational experimental science.

IX. THE NEW SCIENCE OF EXPERIMENTAL PSYCHOLOGY

BRAIN AND MIND

A little over a hundred years ago a reform movement was afoot in the world in the interests of the insane. As was fitting, the movement showed itself first in America, where these unfortunates were humanely cared for at a time when their treatment elsewhere was worse than brutal; but England and France quickly fell into line. The leader on this side of the water was the famous Philadelphian, Dr. Benjamin Rush, "the Sydenham of America"; in England, Dr. William Tuke inaugurated the movement; and in France, Dr. Philippe Pinel, single–handed, led the way. Moved by a common spirit, though acting quite independently, these men raised a revolt against the traditional custom which, spurning the insane as demon–haunted outcasts, had condemned these unfortunates to dungeons, chains, and the lash. Hitherto few people had thought it other than the natural course of events that the "maniac" should be thrust into a dungeon, and perhaps chained to the wall with the aid of an iron band riveted permanently about his neck or waist. Many an unfortunate, thus manacled, was held to the narrow limits of his chain for years together in a cell to which full daylight never penetrated; sometimes—iron being expensive—the chain was so short that the wretched victim could not rise to the upright posture or even shift his position upon his squalid pallet of straw.

In America, indeed, there being no Middle Age precedents to crystallize into established customs, the treatment accorded the insane had seldom or never sunk to this level. Partly for this reason, perhaps, the work of Dr. Rush at the Philadelphia Hospital, in 1784, by means of which the insane came to be humanely treated, even to the extent of banishing the lash, has been but little noted, while the work of the European leaders, though belonging to later decades, has been made famous. And perhaps this is not as unjust as it seems, for the step which Rush took, from relatively bad to good, was a far easier one to take than the leap from atrocities to good treatment which the European reformers were obliged to compass. In Paris, for example, Pinel was obliged to ask permission of the authorities even to make the attempt at liberating the insane from their chains, and, notwithstanding his recognized position as a leader of science, he gained but grudging assent, and was regarded as being himself little better than a lunatic for making so manifestly unwise and hopeless an attempt. Once the attempt had been made, however, and carried to a successful issue, the amelioration wrought in the condition of the insane was so patent that the fame of Pinel's work at the Bicetre and the Salpetriere went abroad apace. It required, indeed, many years to complete it in Paris, and a lifetime of effort on the part of Pinel's pupil Esquirol and others to extend the reform to the provinces; but the epochal turning–point had been reached with Pinel's labors of the closing years of the eighteenth century.

The significance of this wise and humane reform, in the present connection, is the fact that these studies of the insane gave emphasis to the novel idea, which by–and–by became accepted as beyond question, that "demoniacal possession" is in reality no more than the outward expression of a diseased condition of the brain. This realization made it clear, as never before, how intimately the mind and the body are linked one to the other. And so it chanced that, in striking the shackles from the insane, Pinel and his confreres struck a blow also, unwittingly, at time–honored philosophical traditions. The liberation of the insane from their dungeons was an augury of the liberation of psychology from the musty recesses of metaphysics. Hitherto psychology, in so far as it existed at all, was but the subjective study of individual minds; in future it must become objective as well, taking into account also the relations which the mind bears to the body, and in particular to the brain and nervous system.

The necessity for this collocation was advocated quite as earnestly, and even more directly, by another worker of this period, whose studies were allied to those of alienists, and who, even more actively than they, focalized his attention upon the brain and its functions. This earliest of specialists in brain studies was a German by birth but Parisian by adoption, Dr. Franz Joseph Gall, originator of the since-notorious system of phrenology. The merited disrepute into which this system has fallen through the exposition of peripatetic charlatans should not make us forget that Dr. Gall himself was apparently a highly educated physician, a careful student of the brain and mind according to the best light of his time, and, withal, an earnest and honest believer in the validity of the system he had originated. The system itself, taken as a whole, was hopelessly faulty, yet it was not without its latent germ of truth, as later studies were to show. How firmly its author himself believed in it is evidenced by the paper which he contributed to the French Academy of Sciences in 1808. The paper itself was referred to a committee of which Pinel and Cuvier were members. The verdict of this committee was adverse, and justly so; yet the system condemned had at least one merit which its detractors failed to realize. It popularized the conception that the brain is the organ of mind. Moreover, by its insistence it rallied about it a band of scientific supporters, chief of whom was Dr. Kaspar Spurzlieim, a man of no mean abilities, who became the propagandist of phrenology in England and in America. Of course such advocacy and popularity stimulated opposition as well, and out of the disputations thus arising there grew presently a general interest in the brain as the organ of mind, quite aside from any preconceptions whatever as to the doctrines of Gall and Spurzheim.

Prominent among the unprejudiced class of workers who now appeared was the brilliant young Frenchman Louis Antoine Desmoulins, who studied first under the tutorage of the famous Magendie, and published jointly with him a classical work on the nervous system of vertebrates in 1825. Desmoulins made at least one discovery of epochal importance. He observed that the brains of persons dying in old age were lighter than the average and gave visible evidence of atrophy, and he reasoned that such decay is a normal accompaniment of senility. No one nowadays would question the accuracy of this observation, but the scientific world was not quite ready for it in 1825; for when Desmoulins announced his discovery to the French Academy, that august and somewhat patriarchal body was moved to quite unscientific wrath, and forbade the young iconoclast the privilege of further hearings. From which it is evident that the partially liberated spirit of the new psychology had by no means freed itself altogether, at the close of the first quarter of the nineteenth century, from the metaphysical cobwebs of its long incarceration.

FUNCTIONS OF THE NERVES

While studies of the brain were thus being inaugurated, the nervous system, which is the channel of communication between the brain and the outside world, was being interrogated with even more tangible results. The inaugural discovery was made in 1811 by Dr. (afterwards Sir Charles) Bell,[1] the famous English surgeon and experimental physiologist. It consisted of the observation that the anterior roots of the spinal nerves are given over to the function of conveying motor impulses from the brain outward, whereas the posterior roots convey solely sensory impulses to the brain from without. Hitherto it had been supposed that all nerves have a similar function, and the peculiar distribution of the spinal nerves had been an unsolved puzzle.

Bell's discovery was epochal; but its full significance was not appreciated for a decade, nor, indeed, was its validity at first admitted. In Paris, in particular, then the court of final appeal in all matters scientific, the alleged discovery was looked at askance, or quite ignored. But in 1823 the subject was taken up by the recognized leader of French physiology—Francois Magendie—in the course of his comprehensive experimental studies of the nervous system, and Bell's conclusions were subjected to the most rigid experimental tests and found altogether valid. Bell himself, meanwhile, had turned his attention to the cranial nerves, and had proved that these also are divisible into two sets—sensory and motor. Sometimes, indeed, the two sets of filaments are combined into one nerve cord, but if traced to their origin these are found to arise from different brain centres. Thus it was clear that a hitherto unrecognized duality of function pertains to the entire extra-cranial nervous system. Any impulse sent from the periphery to the brain must be conveyed along a perfectly definite channel; the response from the brain, sent out to the peripheral muscles, must traverse an equally definite and altogether different course. If either channel is interrupted—as by the section of its particular nerve tract—the corresponding message is denied transmission as effectually as an electric current is stopped by the section of the transmitting wire.

Experimenters everywhere soon confirmed the observations of Bell and Magendie, and, as always happens after a great discovery, a fresh impulse was given to investigations in allied fields. Nevertheless, a full decade elapsed before another discovery of comparable importance was made. Then Marshall Hall, the most famous of English physicians of his day, made his classical observations on the phenomena that henceforth were to be known as reflex action. In 1832, while experimenting one day with a decapitated newt, he observed that the headless creature's limbs would contract in direct response to certain stimuli. Such a response could no longer be secured if the spinal nerves supplying a part were severed. Hence it was clear that responsive centres exist in the spinal cord capable of receiving a sensory message and of transmitting a motor impulse in reply—a function hitherto supposed to be reserved for the brain. Further studies went to show that such phenomena of reflex action on the part of centres lying outside the range of consciousness, both in the spinal cord and in the brain itself, are extremely common; that, in short, they enter constantly into the activities of every living organism and have a most important share in the sum total of vital movements. Hence, Hall's discovery must always stand as one of the great mile—stones of the advance of neurological science.

Hall gave an admirably clear and interesting account of his experiments and conclusions in a paper before the Royal Society, "On the Reflex Functions of the Medulla Oblongata and the Medulla Spinalis," from which, as published in the Transactions of the society for 1833, we may quote at some length:

"In the entire animal, sensation and voluntary motion, functions of the cerebrum, combine with the functions of the medulla oblongata and medulla spinalis, and may therefore render it difficult or impossible to determine those which are peculiar to each; if, in an animal deprived of the brain, the spinal marrow or the nerves supplying the muscles be stimulated, those muscles, whether voluntary or respiratory, are equally thrown into contraction, and, it may be added, equally in the complete and in the mutilated animal; and, in the case of the nerves, equally in limbs connected with and detached from the spinal marrow.

"The operation of all these various causes may be designated centric, as taking place AT, or at least in a direction FROM, central parts of the nervous system. But there is another function the phenomena of which are of a totally different order and obey totally different laws, being excited by causes in a situation which is EXCENTRIC in the nervous system—that is, distant from the nervous centres. This mode of action has not, I think, been hitherto distinctly understood by physiologists.

"Many of the phenomena of this principle of action, as they occur in the limbs, have certainly been observed. But, in the first place, this function is by no means confined to the limbs; for, while it imparts to each muscle its appropriate tone, and to each system of muscles its appropriate equilibrium or balance, it performs the still more important office of presiding over the orifices and terminations of each of the internal canals in the animal economy, giving them their due form and action; and, in the second place, in the instances in which the phenomena of this function have been noticed, they have been confounded, as I have stated, with those of sensation and volition; or, if they have been distinguished from these, they have been too indefinitely denominated instinctive, or automatic. I have been compelled, therefore, to adopt some new designation for them, and I shall now give the reasons for my choice of that which is given in the title of this paper—'Reflex Functions.'

"This property is characterized by being EXCITED in its action and REFLEX in its course: in every instance in which it is exerted an impression made upon the extremities of certain nerves is conveyed to the medulla oblongata or the medulla spinalis, and is reflected along the nerves to parts adjacent to, or remote from, that which has received the impression.

"It is by this reflex character that the function to which I have alluded is to be distinguished from every other. There are, in the animal economy, four modes of muscular action, of muscular contraction. The first is that designated VOLUNTARY: volition, originated in the cerebrum and spontaneous in its acts, extends its influence along the spinal marrow and the motor nerves in a DIRECT LINE to the voluntary muscles. The SECOND is that of RESPIRATION: like volition, the motive influence in respiration passes in a DIRECT LINE from one point of the nervous system to certain muscles; but as voluntary motion seems to originate in the cerebrum, so the respiratory motions originate in the medulla oblongata: like the voluntary motions, the motions of respirations are spontaneous; they continue, at least, after the eighth pair of nerves have been divided. The THIRD kind of muscular action in the animal economy is that termed involuntary: it depends upon the principle of irritability and requires the IMMEDIATE application of a stimulus to the nervo—muscular fibre itself. These three kinds of muscular motion are well known to physiologists; and I believe they are all which have been hitherto pointed out. There is, however, a FOURTH, which subsists, in part, after the voluntary and respiratory motions have ceased, by the removal of the cerebrum and medulla oblongata, and which is attached to the medulla spinalis, ceasing itself when this is removed, and leaving the irritability undiminished. In this kind of muscular motion the motive influence does not originate in any central part of the nervous system, but from a distance from that centre; it is neither spontaneous in its action nor direct in its course; it is, on the contrary, EXCITED by the application of appropriate stimuli, which are not, however, applied immediately to the muscular or nervo–muscular fibre, but to certain membraneous parts, whence the impression is carried through the medulla, REFLECTED and reconducted to the part impressed, or conducted to a part remote from it in which muscular contraction is effected.

"The first three modes of muscular action are known only by actual movements of muscular contractions. But the reflex function exists as a continuous muscular action, as a power presiding over organs not actually in a state of motion, preserving in some, as the glottis, an open, in others, as the sphincters, a closed form, and in the limbs a due degree of equilibrium or balanced muscular action—a function not, I think, hitherto recognized by physiologists.

The three kinds of muscular motion hitherto known may be distinguished in another way. The muscles of voluntary motion and of respiration may be excited by stimulating the nerves which supply them, in any part of their course, whether at their source as a part of the medulla oblongata or the medulla spinalis or exterior to the spinal canal: the muscles of involuntary motion are chiefly excited by the actual contact of stimuli. In the case of the reflex function alone the muscles are excited by a stimulus acting mediately and indirectly in a curved and reflex course, along superficial subcutaneous or submucous nerves proceeding from the medulla. The first three of these causes of muscular motion may act on detached limbs or muscles. The last requires the connection with the medulla to be preserved entire.

"All the kinds of muscular motion may be unduly excited, but the reflex function is peculiar in being excitable in two modes of action, not previously subsisting in the animal economy, as in the case of sneezing, coughing, vomiting, etc. The reflex function also admits of being permanently diminished or augmented and of taking on some other morbid forms, of which I shall treat hereafter.

"Before I proceed to the details of the experiments upon which this disposition rests, it may be well to point out several instances in illustration of the various sources of and the modes of muscular action which have been enumerated. None can be more familiar than the act of swallowing. Yet how complicated is the act! The apprehension of the food by the teeth and tongue, etc., is voluntary, and cannot, therefore, take place in an animal from which the cerebrum is removed. The transition of food over the glottis and along the middle and lower part of the pharynx depends upon the reflex action: it can take place in animals from which the cerebrum has been removed or the ninth pair of nerves divided; but it requires the connection with the medulla oblongata to be preserved entirely; and the actual contact of some substance which may act as a stimulus: it is attended by the accurate closure of the glottis and by the contraction of the pharynx. The completion of the act of deglutition is dependent upon the stimulus immediately impressed upon the muscular fibre of the oesophagus, and is the result of excited irritability.

"However plain these observations may have made the fact that there is a function of the nervous muscular system distinct from sensation, from the voluntary and respiratory motions, and from irritability, it is right, in every such inquiry as the present, that the statements and reasonings should be made with the experiment, as it were, actually before us. It has already been remarked that the voluntary and respiratory motions are spontaneous, not necessarily requiring the agency of a stimulus. If, then, an animal can be placed in such circumstances that such motions will certainly not take place, the power of moving remaining, it may be concluded that volition and the motive influence of respiration are annihilated. Now this is effected by removing the cerebrum and the medulla oblongata. These facts are fully proved by the experiments of Legallois and M. Flourens, and by several which I proceed to detail, for the sake of the opportunity afforded by doing so of stating the arguments most clearly.

"I divided the spinal marrow of a very lively snake between the second and third vertebrae. The movements of the animal were immediately before extremely vigorous and unintermitted. From the moment of the division of the spinal marrow it lay perfectly tranquil and motionless, with the exception of occasional gaspings and slight movements of the head. It became quite evident that this state of quiescence would continue indefinitely were the animal secured from all external impressions.

"Being now stimulated, the body began to move with great activity, and continued to do so for a considerable time, each change of position or situation bringing some fresh part of the surface of the animal into contact with the table or other objects and renewing the application of stimulants.

"At length the animal became again quiescent; and being carefully protected from all external impressions it moved no more, but died in the precise position and form which it had last assumed.

"It requires a little manoeuvre to perform this experiment successfully: the motions of the animal must be watched and slowly and cautiously arrested by opposing some soft substance, as a glove or cotton wool; they are by this means gradually lulled into quiescence. The slightest touch with a hard substance, the slightest stimulus, will, on the other hand, renew the movements on the animal in an active form. But that this phenomenon does not depend upon sensation is further fully proved by the facts that the position last assumed, and the stimuli, may be such as would be attended by extreme or continued pain, if the sensibility were undestroyed: in one case the animal remained partially suspended over the acute edge of the table; in others the infliction of punctures and the application of a lighted taper did not prevent the animal, still possessed of active powers of motion, from passing into a state of complete and permanent quiescence."

In summing up this long paper Hall concludes with this sentence: "The reflex function appears in a word to be the COMPLEMENT of the functions of the nervous system hitherto known."[2]

All these considerations as to nerve currents and nerve tracts becoming stock knowledge of science, it was natural that interest should become stimulated as to the exact character of these nerve tracts in themselves, and all the more natural in that the perfected microscope was just now claiming all fields for its own. A troop of observers soon entered upon the study of the nerves, and the leader here, as in so many other lines of microscopical research, was no other than Theodor Schwann. Through his efforts, and with the invaluable aid of such other workers as Remak, Purkinje, Henle, Muller, and the rest, all the mystery as to the general characteristics of nerve tracts was cleared away. It came to be known that in its essentials a nerve tract is a tenuous fibre or thread of protoplasm stretching between two terminal points in the organism, one of such termini being usually a cell of the brain or spinal cord, the other a distribution–point at or near the periphery—for example, in a muscle or in the skin. Such a fibril may have about it a protective covering, which is known as the sheath of Schwann; but the fibril itself is the essential nerve tract; and in many cases, as Remak presently discovered, the sheath is dispensed with, particularly in case of the nerves of the so—called sympathetic system.

This sympathetic system of ganglia and nerves, by—the—bye, had long been a puzzle to the physiologists. Its ganglia, the seeming centre of the system, usually minute in size and never very large, are found everywhere through the organism, but in particular are gathered into a long double chain which lies within the body cavity, outside the spinal column, and represents the sole nervous system of the non—vertebrated organisms. Fibrils from these ganglia were seen to join the cranial and spinal nerve fibrils and to accompany them everywhere, but what special function they subserved was long a mere matter of conjecture and led to many absurd speculations. Fact was not substituted for conjecture until about the year 1851, when the great Frenchman Claude Bernard conclusively proved that at least one chief function of the sympathetic fibrils is to cause contraction of the walls of the arterioles of the system, thus regulating the blood—supply of any given part. Ten years earlier Henle had demonstrated the existence of annular bands of muscle fibres in the arterioles, hitherto a much—mooted question, and several tentative explanations of the action of these fibres had been made, particularly by the brothers Weber, by Stilling, who, as early as 1840, had ventured to speak of "vaso—motor" nerves, and by Schiff, who was hard upon the same track at the time of Bernard's discovery. But a clear light was not thrown on the subject until Bernard's experiments were made in 1851. The experiments were soon after confirmed and extended by Brown—Sequard, Waller, Budge, and numerous others, and henceforth physiologists felt that they understood how the blood—supply of any given part is regulated by the nervous system.

In reality, however, they had learned only half the story, as Bernard himself proved only a few years later by opening up a new and quite unsuspected chapter. While experimenting in 1858 he discovered that there are certain nerves supplying the heart which, if stimulated, cause that organ to relax and cease beating. As the heart is essentially nothing more than an aggregation of muscles, this phenomenon was utterly puzzling and without precedent in the experience of physiologists. An impulse travelling along a motor nerve had been supposed to be able to cause a muscular contraction and to do nothing else; yet here such an impulse had exactly the opposite effect. The only tenable explanation seemed to be that this particular impulse must arrest or inhibit the action of the impulses that ordinarily cause the heart muscles to contract. But the idea of such inhibition of one impulse by another was utterly novel and at first difficult to comprehend. Gradually, however, the idea took its place in the current knowledge of nerve physiology, and in time it came to be understood that what happens in the case of the heart nerve–supply is only a particular case under a very general, indeed universal, form of nervous action. Growing out of Bernard's initial discovery came the final understanding that the entire nervous system is a mechanism of centres subordinate and centres superior, the action of the one of which may be counteracted and annulled in effect by the action of the other. This applies not merely to such physical processes as heart–beats and arterial contraction and relaxing, but to the most intricate functionings which have their counterpart in psychical processes as well. Thus the observation of the inhibition of the heart's action by a nervous impulse furnished the point of departure for studies that led to a better understanding of the modus operandi of the mind's activities than had ever previously been attained by the most subtle of psychologists.

PSYCHO–PHYSICS

The work of the nerve physiologists had thus an important bearing on questions of the mind. But there was another company of workers of this period who made an even more direct assault upon the "citadel of thought." A remarkable school of workers had been developed in Germany, the leaders being men who, having more or less of innate metaphysical bias as a national birthright, had also the instincts of the empirical scientist, and whose educational equipment included a profound knowledge not alone of physiology and psychology, but of physics and mathematics as well. These men undertook the novel task of interrogating the relations of body and mind from the standpoint of physics. They sought to apply the vernier and the balance, as far as might be, to the intangible processes of mind.

The movement had its precursory stages in the early part of the century, notably in the mathematical psychology of Herbart, but its first definite output to attract general attention came from the master–hand of Hermann Helmholtz in 1851. It consisted of the accurate measurement of the speed of transit of a nervous impulse along a nerve tract. To make such measurement had been regarded as impossible, it being supposed that the flight of the nervous impulse was practically instantaneous. But Helmholtz readily demonstrated the contrary, showing that the nerve cord is a relatively sluggish message–bearer. According to his experiments, first performed upon the frog, the nervous "current" travels less than one hundred feet per second. Other experiments performed soon afterwards by Helmholtz himself, and by various followers, chief among whom was Du Bois–Reymond, modified somewhat the exact figures at first obtained, but did not change the general bearings of the early results. Thus the nervous impulse was shown to be something far different, as regards speed of transit, at any rate, from the electric current to which it had been so often likened. An electric current would flash halfway round the globe while a nervous impulse could travel the length of the human body—from a man's foot to his brain.

The tendency to bridge the gulf that hitherto had separated the physical from the psychical world was further evidenced in the following decade by Helmholtz's remarkable but highly technical study of the sensations of sound and of color in connection with their physical causes, in the course of which he revived the doctrine of color vision which that other great physiologist and physicist, Thomas Young, had advanced half a century before. The same tendency was further evidenced by the appearance, in 1852, of Dr. Hermann Lotze's famous Medizinische Psychologie, oder Physiologie der Seele, with its challenge of the old myth of a "vital force." But the most definite expression of the new movement was signalized in 1860, when Gustav Fechner published his classical work called Psychophysik. That title introduced a new word into the vocabulary of science. Fechner explained it by saying, "I mean by psychophysics an exact theory of the relation between spirit and body, and, in a general way, between the physical and the psychic worlds." The title became famous and the brunt of many a controversy. So also did another phrase which Fechner introduced in the course of his book—the phrase "physiological psychology." In making that happy collocation of words Fechner virtually christened a new science.

FECHNER EXPOUNDS WEBER'S LAW

The chief purport of this classical book of the German psycho–physiologist was the elaboration and explication of experiments based on a method introduced more than twenty years earlier by his countryman E. H. Weber, but which hitherto had failed to attract the attention it deserved. The method consisted of the measurement and analysis of the definite relation existing between external stimuli of varying degrees of intensity (various sounds, for example) and the mental states they induce. Weber's experiments grew out of the familiar observation that the nicety of our discriminations of various sounds, weights, or visual images depends upon the magnitude of each particular cause of a sensation in its relation with other similar causes. Thus, for example, we cannot see the stars in the daytime, though they shine as brightly then as at night. Again, we seldom notice the ticking of a clock in the daytime, though it may become almost painfully audible in the silence of the night. Yet again, the difference between an ounce weight and a two–ounce weight is clearly enough appreciable when we lift the two, but one cannot discriminate in the same way between a five–pound weight and a weight of one ounce over five pounds.

This last example, and similar ones for the other senses, gave Weber the clew to his novel experiments. Reflection upon every–day experiences made it clear to him that whenever we consider two visual sensations, or two auditory sensations, or two sensations of weight, in comparison one with another, there is always a limit to the keenness of our discrimination, and that this degree of keenness varies, as in the case of the weights just cited, with the magnitude of the exciting cause.

Weber determined to see whether these common experiences could be brought within the pale of a general law. His method consisted of making long series of experiments aimed at the determination, in each case, of what came to be spoken of as the least observable difference between the stimuli. Thus if one holds an ounce weight in each hand, and has tiny weights added to one of them, grain by grain, one does not at first perceive a difference; but presently, on the addition of a certain grain, he does become aware of the difference. Noting now how many grains have been added to produce this effect, we have the weight which represents the least appreciable difference when the standard is one ounce.

Now repeat the experiment, but let the weights be each of five pounds. Clearly in this case we shall be obliged to add not grains, but drachms, before a difference between the two heavy weights is perceived. But whatever the exact amount added, that amount represents the stimulus producing a just–perceivable sensation of difference when the standard is five pounds. And so on for indefinite series of weights of varying magnitudes. Now came Weber's curious discovery. Not only did he find that in repeated experiments with the same pair of weights the measure of "just–{p}erceivable difference" remained approximately fixed, but he found, further, that a remarkable fixed relation exists between the stimuli of different magnitude. If, for example, he had found it necessary, in the case of the ounce weights, to add one–fiftieth of an ounce to the one before a difference was detected, he found also, in the case of the five–pound weights, that one–fiftieth of five pounds must be added before producing the same result. And so of all other weights; the amount added to produce the stimulus of "least–appreciable difference" always bore the same mathematical relation to the magnitude of the weight used, be that magnitude great or small.

Weber found that the same thing holds good for the stimuli of the sensations of sight and of hearing, the differential stimulus bearing always a fixed ratio to the total magnitude of the stimuli. Here, then, was the law he had sought.

Weber's results were definite enough and striking enough, yet they failed to attract any considerable measure of attention until they were revived and extended by Fechner and brought before the world in the famous work on psycho–physics. Then they precipitated a veritable melee. Fechner had not alone verified the earlier results (with certain limitations not essential to the present consideration), but had invented new methods of making similar tests, and had reduced the whole question to mathematical treatment. He pronounced Weber's discovery the fundamental law of psycho–physics. In honor of the discoverer, he christened it Weber's Law. He clothed the law in words and in mathematical formulae, and, so to say, launched it full tilt at the heads of the psychological world. It made a fine commotion, be assured, for it was the first widely heralded bulletin of the new psychology in its march upon the strongholds of the time–honored metaphysics. The accomplishments of the microscopists and the nerve physiologists had been but preliminary—mere border skirmishes of uncertain import. But here was proof that the iconoclastic movement meant to invade the very heart of the sacred territory of mind—a territory from which tangible objective fact had been supposed to be forever barred.

PHYSIOLOGICAL PSYCHOLOGY

Hardly had the alarm been sounded, however, before a new movement was made. While Fechner's book was fresh from the press, steps were being taken to extend the methods of the physicist in yet another way to the intimate processes of the mind. As Helmholtz had shown the rate of nervous impulsion along the nerve tract to be measurable, it was now sought to measure also the time required for the central nervous mechanism to perform its work of receiving a message and sending out a response. This was coming down to the very threshold of mind. The attempt was first made by Professor Donders in 1861, but definitive results were only obtained after many years of experiment on the part of a host of observers. The chief of these, and the man who has stood in the forefront of the new movement and has been its recognized leader throughout the remainder of the century, is Dr. Wilhelm Wundt, of Leipzig.

The task was not easy, but, in the long run, it was accomplished. Not alone was it shown that the nerve centre requires a measurable time for its operations, but much was learned as to conditions that modify this time. Thus it was found that different persons vary in the rate of their central nervous activity—which explained the "personal equation" that the astronomer Bessel had noted a half–century before. It was found, too, that the rate of activity varies also for the same person under different conditions, becoming retarded, for example, under influence of fatigue, or in case of certain diseases of the brain. All details aside, the essential fact emerges, as an experimental demonstration, that the intellectual processes—sensation, apperception, volition—are linked irrevocably with the activities of the central nervous tissues, and that these activities, like all other physical processes, have a time element. To that old school of psychologists, who scarcely cared more for the human head than for the heels—being interested only in the mind—such a linking of mind and body as was thus demonstrated was naturally disquieting. But whatever the inferences, there was no escaping the facts.

Of course this new movement has not been confined to Germany. Indeed, it had long had exponents elsewhere. Thus in England, a full century earlier, Dr. Hartley had championed the theory of the close and indissoluble dependence of the mind upon the brain, and formulated a famous vibration theory of association that still merits careful consideration. Then, too, in France, at the beginning of the century, there was Dr. Cabanis with his tangible, if crudely phrased, doctrine that the brain digests impressions and secretes thought as the stomach digests food and the liver secretes bile. Moreover, Herbert Spencer's Principles of Psychology, with its avowed co–ordination of mind and body and its vitalizing theory of evolution, appeared in 1855, half a decade before the work of Fechner. But these influences, though of vast educational value, were theoretical rather than demonstrative, and the fact remains that the experimental work which first attempted to gauge mental operations by physical principles was mainly done in Germany. Wundt's Physiological Psychology, with its full preliminary descriptions of the anatomy of the nervous system, gave tangible expression to the growth of the new movement in 1874; and four years later, with the opening of his laboratory of physiological psychology at the University of Leipzig, the new psychology may be said to have gained a permanent foothold and to have forced itself into official recognition. From then on its conquest of the world was but a matter of time.

It should be noted, however, that there is one other method of strictly experimental examination of the mental field, latterly much in vogue, which had a different origin.

This is the scientific investigation of the phenomena of hypnotism. This subject was rescued from the hands of charlatans, rechristened, and subjected to accurate investigation by Dr. James Braid, of Manchester, as early as 1841. But his results, after attracting momentary attention, fell from view, and, despite desultory efforts, the subject was not again accorded a general hearing from the scientific world until 1878, when Dr. Charcot took it up at the Salpetriere, in Paris, followed soon afterwards by Dr. Rudolf Heidenhain, of Breslau, and a host of other experimenters. The value of the method in the study of mental states was soon apparent. Most of Braid's experiments were repeated, and in the main his results were confirmed. His explanation of hypnotism, or artificial somnambulism, as a self–induced state, independent of any occult or supersensible influence, soon gained general credence. His belief that the initial stages are due to fatigue of nervous centres, usually from excessive stimulation, has not been supplanted, though supplemented by notions growing out of the new knowledge as to subconscious mentality in general, and the inhibitory influence of one centre over another in the central nervous mechanism.

THE BRAIN AS THE ORGAN OF MIND

These studies of the psychologists and pathologists bring the relations of mind and body into sharp relief. But even more definite in this regard was the work of the brain physiologists. Chief of these, during the middle period of the century, was the man who is sometimes spoken of as the "father of brain physiology," Marie Jean Pierre Flourens, of the Jardin des Plantes of Paris, the pupil and worthy successor of Magendie. His experiments in nerve physiology were begun in the first quarter of the century, but his local experiments upon the brain itself were not culminated until about 1842. At this time the old dispute over phrenology had broken out afresh, and the studies of Flourens were aimed, in part at least, at the strictly scientific investigation of this troublesome topic.

In the course of these studies Flourens discovered that in the medulla oblongata, the part of the brain which connects that organ with the spinal cord, there is a centre of minute size which cannot be injured in the least without causing the instant death of the animal operated upon. It may be added that it is this spot which is reached by the needle of the garroter in Spanish executions, and that the same centre also is destroyed when a criminal is "successfully" hanged, this time by the forced intrusion of a process of the second cervical vertebra. Flourens named this spot the "vital knot." Its extreme importance, as is now understood, is due to the fact that it is the centre of nerves that supply the heart; but this simple explanation, annulling the conception of a specific "life centre," was not at once apparent.

Other experiments of Flourens seemed to show that the cerebellum is the seat of the centres that co—ordinate muscular activities, and that the higher intellectual faculties are relegated to the cerebrum. But beyond this, as regards localization, experiment faltered. Negative results, as regards specific faculties, were obtained from all localized irritations of the cerebrum, and Flourens was forced to conclude that the cerebral lobe, while being undoubtedly the seat of higher intellection, performs its functions with its entire structure. This conclusion, which incidentally gave a quietus to phrenology, was accepted generally, and became the stock doctrine of cerebral physiology for a generation.

It will be seen, however, that these studies of Flourens had a double bearing. They denied localization of cerebral functions, but they demonstrated the localization of certain nervous processes in other portions of the brain. On the whole, then, they spoke positively for the principle of localization of function in the brain, for which a certain number of students contended; while their evidence against cerebral localization was only negative. There was here and there an observer who felt that this negative testimony was not conclusive. In particular, the German anatomist Meynert, who had studied the disposition of nerve tracts in the cerebrum, was led to believe that the anterior portions of the cerebrum must have motor functions in preponderance; the posterior positions, sensory functions. Somewhat similar conclusions were reached also by Dr. Hughlings—Jackson, in England, from his studies of epilepsy. But no positive evidence was forthcoming until 1861, when Dr. Paul Broca brought before the Academy of Medicine in Paris a case of brain lesion which he regarded as having most important bearings on the question of cerebral localization.

The case was that of a patient at the Bicetre, who for twenty years had been deprived of the power of speech, seemingly through loss of memory of words. In 1861 this patient died, and an autopsy revealed that a certain convolution of the left frontal lobe of his cerebrum had been totally destroyed by disease, the remainder of his brain being intact. Broca felt that this observation pointed strongly to a localization of the memory of words in a definite area of the brain. Moreover, it transpired that the case was not without precedent. As long ago as 1825 Dr. Boillard had been led, through pathological studies, to locate definitely a centre for the articulation of words in the frontal lobe, and here and there other observers had made tentatives in the same direction. Boillard had even followed the matter up with pertinacity, but the world was not ready to listen to him. Now, however, in the half–decade that followed Broca's announcements, interest rose to fever–beat, and through the efforts of Broca, Boillard, and numerous others it was proved that a veritable centre having a strange domination over the memory of articulate words has its seat in the third convolution of the frontal lobe of the cerebrum, usually in the left hemisphere. That part of the brain has since been known to the English–speaking world as the convolution of Broca, a name which, strangely enough, the discoverer's compatriots have been slow to accept.

This discovery very naturally reopened the entire subject of brain localization. It was but a short step to the inference that there must be other definite centres worth the seeking, and various observers set about searching for them. In 1867 a clew was gained by Eckhard, who, repeating a forgotten experiment by Haller and Zinn of the previous century, removed portions of the brain cortex of animals, with the result of producing convulsions. But the really vital departure was made in 1870 by the German investigators Fritsch and Hitzig, who, by stimulating definite areas of the cortex of animals with a galvanic current, produced contraction of definite sets of muscles of the opposite side of the body. These most important experiments, received at first with incredulity, were repeated and extended in 1873 by Dr. David Ferrier, of London, and soon afterwards by a small army of independent workers everywhere, prominent among whom were Franck and Pitres in France, Munck and Goltz in Germany, and Horsley and Schafer in England. The detailed results, naturally enough, were not at first all in harmony. Some observers, as Goltz, even denied the validity of the conclusions in toto. But a consensus of opinion, based on multitudes of experiments, soon placed the broad general facts for which Fritsch and Hitzig contended beyond controversy. It was found, indeed, that the cerebral centres of motor activities have not quite the finality at first ascribed to them by some observers, since it may often happen that after the destruction of a centre, with attending loss of function, there may be a gradual restoration of the lost function, proving that other centres have acquired the capacity to take the place of the one destroyed. There are limits to this capacity for substitution, however, and with this qualification the definiteness of the localization of motor functions in the cerebral cortex has become an accepted part of brain physiology.

Nor is such localization confined to motor centres. Later experiments, particularly of Ferrier and of Munck, proved that the centres of vision are equally restricted in their location, this time in the posterior lobes of the brain, and that hearing has likewise its local habitation. Indeed, there is every reason to believe that each form of primary sensation is based on impressions which mainly come to a definitely localized goal in the brain. But all this, be it understood, has no reference to the higher forms of intellection. All experiment has proved futile to localize these functions, except indeed to the extent of corroborating the familiar fact of their dependence upon the brain, and, somewhat problematically, upon the anterior lobes of the cerebrum in particular. But this is precisely what should be expected, for the clearer insight into the nature of mental processes makes it plain that in the main these alleged "faculties" are not in themselves localized. Thus, for example, the "faculty" of language is associated irrevocably with centres of vision, of hearing, and of muscular activity, to go no further, and only becomes possible through the association of these widely separated centres. The destruction of Broca's centre, as was early discovered, does not altogether deprive a patient of his knowledge of language. He may be totally unable to speak (though as to this there are all degrees of variation), and yet may comprehend what is said to him, and be able to read, think, and even write correctly. Thus it appears that Broca's centre is peculiarly bound up with the capacity for articulate speech, but is far enough from being the seat of the faculty of language in its entirety.

In a similar way, most of the supposed isolated "faculties" of higher intellection appear, upon clearer analysis, as complex aggregations of primary sensations, and hence necessarily dependent upon numerous and scattered centres. Some "faculties," as memory and volition, may be said in a sense to be primordial endowments of every nerve cell—even of every body cell. Indeed, an ultimate analysis relegates all intellection, in its primordial adumbrations, to every particle of living matter. But such refinements of analysis, after all, cannot hide the fact that certain forms of higher intellection involve a pretty definite collocation and elaboration of special sensations. Such specialization, indeed, seems a necessary accompaniment of mental evolution. That every such specialized function has its localized centres of co-ordination, of some such significance as the demonstrated centres of articulate speech, can hardly be in doubt—though this, be it understood, is an induction, not as yet a demonstration. In other words, there is every reason to believe that numerous "centres," in this restricted sense, exist in the brain that have as yet eluded the investigator. Indeed, the current conception regards the entire cerebral cortex as chiefly composed of centres of ultimate co-ordination of impressions, which in their cruder form are received by more primitive nervous tissues—the basal ganglia, the cerebellum and medulla, and the spinal cord.

This, of course, is equivalent to postulating the cerebral cortex as the exclusive seat of higher intellection. This proposition, however, to which a safe induction seems to lead, is far afield from the substantiation of the old conception of brain localization, which was based on faulty psychology and equally faulty inductions from few premises. The details of Gall's system, as propounded by generations of his mostly unworthy followers, lie quite beyond the pale of scientific discussion. Yet, as I have said, a germ of truth was there—the idea of specialization of cerebral functions—and modern investigators have rescued that central conception from the phrenological rubbish heap in which its discoverer unfortunately left it buried.

THE MINUTE STRUCTURE OF THE BRAIN

The common ground of all these various lines of investigations of pathologist, anatomist, physiologist, physicist, and psychologist is, clearly, the central nervous system—the spinal cord and the brain. The importance of these structures as the foci of nervous and mental activities has been recognized more and more with each new accretion of knowledge, and the efforts to fathom the secrets of their intimate structure has been unceasing. For the earlier students, only the crude methods of gross dissections and microscopical inspection were available. These could reveal something, but of course the inner secrets were for the keener insight of the microscopist alone. And even for him the task of investigation was far from facile, for the central nervous tissues are the most delicate and fragile, and on many accounts the most difficult of manipulation of any in the body.

Special methods, therefore, were needed for this essay, and brain histology has progressed by fitful impulses, each forward jet marking the introduction of some ingenious improvement of mechanical technique, which placed a new weapon in the hands of the investigators.

The very beginning was made in 1824 by Rolando, who first thought of cutting chemically hardened pieces of brain tissues into thin sections for microscopical examination—the basal structure upon which almost all the later advances have been conducted. Muller presently discovered that bichromate of potassium in solution makes the best of fluids for the preliminary preservation and hardening of the tissues. Stilling, in 1842, perfected the method by introducing the custom of cutting a series of consecutive sections of the same tissue, in order to trace nerve tracts and establish spacial relations.

Then from time to time mechanical ingenuity added fresh details of improvement. It was found that pieces of hardened tissue of extreme delicacy can be made better subject to manipulation by being impregnated with collodion or celloidine and embedded in paraffine. Latterly it has become usual to cut sections also from fresh tissues, unchanged by chemicals, by freezing them suddenly with vaporized ether or, better, carbonic acid. By these methods, and with the aid of perfected microtomes, the worker of recent periods avails himself of sections of brain tissues of a tenuousness which the early investigators could not approach.

But more important even than the cutting of thin sections is the process of making the different parts of the section visible, one tissue differentiated from another. The thin section, as the early workers examined it, was practically colorless, and even the crudest details of its structure were made out with extreme difficulty. Remak did, indeed, manage to discover that the brain tissue is cellular, as early as 1833, and Ehrenberg in the same year saw that it is also fibrillar, but beyond this no great advance was made until 1858, when a sudden impulse was received from a new process introduced by Gerlach. The process itself was most simple, consisting essentially of nothing more than the treatment of a microscopical section with a solution of carmine. But the result was wonderful, for when such a section was placed under the lens it no longer appeared homogeneous. Sprinkled through its substance were seen irregular bodies that had taken on a beautiful color, while the matrix in which they were embedded remained unstained. In a word, the central nerve cell had sprung suddenly into clear view.

A most interesting body it proved, this nerve cell, or ganglion cell, as it came to be called. It was seen to be exceedingly minute in size, requiring high powers of the microscope to make it visible. It exists in almost infinite numbers, not, however, scattered at random through the brain and spinal cord. On the contrary, it is confined to those portions of the central nervous masses which to the naked eye appear gray in color, being altogether wanting in the white substance which makes up the chief mass of the brain. Even in the gray matter, though sometimes thickly distributed, the ganglion cells are never in actual contact one with another; they always lie embedded in intercellular tissues, which came to be known, following Virchow, as the neuroglia.

Each ganglion cell was seen to be irregular in contour, and to have jutting out from it two sets of minute fibres, one set relatively short, indefinitely numerous, and branching in every direction; the other set limited in number, sometimes even single, and starting out directly from the cell as if bent on a longer journey. The numerous filaments came to be known as protoplasmic processes; the other fibre was named, after its discoverer, the axis cylinder of Deiters. It was a natural inference, though not clearly demonstrable in the sections, that these filamentous processes are the connecting links between the different nerve cells and also the channels of communication between nerve cells and the periphery of the body. The white substance of brain and cord, apparently, is made up of such connecting fibres, thus bringing the different ganglion cells everywhere into communication one with another.

In the attempt to trace the connecting nerve tracts through this white substance by either macroscopical or microscopical methods, most important aid is given by a method originated by Waller in 1852. Earlier than that, in 1839, Nasse had discovered that a severed nerve cord degenerates in its peripheral portions. Waller discovered that every nerve fibre, sensory or motor, has a nerve cell to or from which it leads, which dominates its nutrition, so that it can only retain its vitality while its connection with that cell is intact. Such cells he named trophic centres. Certain cells of the anterior part of the spinal cord, for example, are the trophic centres of the spinal motor nerves. Other trophic centres, governing nerve tracts in the spinal cord itself, are in the various regions of the brain. It occurred to Waller that by destroying such centres, or by severing the connection at various regions between a nervous tract and its trophic centre, sharply defined tracts could be made to degenerate, and their location could subsequently be accurately defined, as the degenerated tissues take on a changed aspect, both to macroscopical and microscopical observation. Recognition of this principle thus gave the experimenter a new weapon of great efficiency in tracing nervous connections. Moreover, the same principle has wide application in case of the human subject in disease, such as the lesion of nerve tracts or the destruction of centres by localized tumors, by embolisms, or by traumatisms.

All these various methods of anatomical examination combine to make the conclusion almost unavoidable that the central ganglion cells are the veritable "centres" of nervous activity to which so many other lines of research have pointed. The conclusion was strengthened by experiments of the students of motor localization, which showed that the veritable centres of their discovery lie, demonstrably, in the gray cortex of the brain, not in the white matter. But the full proof came from pathology. At the hands of a multitude of observers it was shown that in certain well-known diseases of the spinal cord, with resulting paralysis, it is the ganglion cells themselves that are found to be destroyed.

Similarly, in the case of sufferers from chronic insanities, with marked dementia, the ganglion cells of the cortex of the brain are found to have undergone degeneration. The brains of paretics in particular show such degeneration, in striking correspondence with their mental decadence. The position of the ganglion cell as the ultimate centre of nervous activities was thus placed beyond dispute.

Meantime, general acceptance being given the histological scheme of Gerlach, according to which the mass of the white substance of the brain is a mesh—work of intercellular fibrils, a proximal idea seemed attainable of the way in which the ganglionic activities are correlated, and, through association, built up, so to speak, into the higher mental processes. Such a conception accorded beautifully with the ideas of the associationists, who had now become dominant in psychology. But one standing puzzle attended this otherwise satisfactory correlation of anatomical observations and psychic analyses. It was this: Since, according to the histologist, the intercellular fibres, along which impulses are conveyed, connect each brain cell, directly or indirectly, with every other brain cell in an endless mesh—work, how is it possible that various sets of cells may at times be shut off from one another? Such isolation must take place, for all normal ideation depends for its integrity quite as much upon the shutting—out of the great mass of associations as upon the inclusion of certain other associations. For example, a student in solving a mathematical problem must for the moment become quite oblivious to the special associations that have to do with geography, natural history, and the like. But does histology give any clew to the way in which such isolation may be effected?

Attempts were made to find an answer through consideration of the very peculiar character of the blood—supply in the brain. Here, as nowhere else, the terminal twigs of the arteries are arranged in closed systems, not anastomosing freely with neighboring systems. Clearly, then, a restricted area of the brain may, through the controlling influence of the vasomotor nerves, be flushed with arterial blood while neighboring parts remain relatively anaemic. And since vital activities unquestionably depend in part upon the supply of arterial blood, this peculiar arrangement of the vascular mechanism may very properly be supposed to aid in the localized activities of the central nervous ganglia. But this explanation left much to be desired—in particular when it is recalled that all higher intellection must in all probability involve multitudes of widely scattered centres.

No better explanation was forthcoming, however, until the year 1889, when of a sudden the mystery was cleared away by a fresh discovery. Not long before this the Italian histologist Dr. Camille Golgi had discovered a method of impregnating hardened brain tissues with a solution of nitrate of silver, with the result of staining the nerve cells and their processes almost infinitely better than was possible by the methods of Gerlach, or by any of the multiform methods that other workers had introduced. Now for the first time it became possible to trace the cellular prolongations definitely to their termini, for the finer fibrils had not been rendered visible by any previous method of treatment. Golgi himself proved that the set of fibrils known as protoplasmic prolongations terminate by free extremities, and have no direct connection with any cell save the one from which they spring. He showed also that the axis cylinders give off multitudes of lateral branches not hitherto suspected. But here he paused, missing the real import of the discovery of which he was hard on the track. It remained for the Spanish histologist Dr. S. Ramon y Cajal to follow up the investigation by means of an improved application of Golgi's method of staining, and to demonstrate that the axis cylinders, together with all their collateral branches, though sometimes extending to a great distance, yet finally terminate, like the other cell prolongations, in arborescent fibrils having free extremities. In a word, it was shown that each central nerve cell, with its fibrillar offshoots, is an isolated entity. Instead of being in physical connection with a multitude of other nerve cells, it has no direct physical connection with any other nerve cell whatever.

When Dr. Cajal announced his discovery, in 1889, his revolutionary claims not unnaturally amazed the mass of histologists. There were some few of them, however, who were not quite unprepared for the revelation; in particular His, who had half suspected the independence of the cells, because they seemed to develop from dissociated centres; and Forel, who based a similar suspicion on the fact that he had never been able actually to trace a fibre from one cell to another. These observers then came readily to repeat Cajal's experiments. So also did the veteran histologist Kolliker, and soon afterwards all the leaders everywhere. The result was a practically unanimous confirmation of the Spanish histologist's claims, and within a few months after his announcements the old theory of union of nerve cells into an endless mesh−work was completely discarded, and the theory of isolated nerve elements—the theory of neurons, as it came to be called—was fully established in its place.

As to how these isolated nerve cells functionate, Dr. Cajal gave the clew from the very first, and his explanation has met with universal approval.

In the modified view, the nerve cell retains its old position as the storehouse of nervous energy. Each of the filaments jutting out from the cell is held, as before, to be indeed a transmitter of impulses, but a transmitter that operates intermittently, like a telephone wire that is not always "connected," and, like that wire, the nerve fibril operates by contact and not by continuity. Under proper stimulation the ends of the fibrils reach out, come in contact with other end fibrils of other cells, and conduct their destined impulse. Again they retract, and communication ceases for the time between those particular cells. Meantime, by a different arrangement of the various conductors, different sets of cells are placed in communication, different associations of nervous impulses induced, different trains of thought engendered. Each fibril when retracted becomes a non–conductor, but when extended and in contact with another fibril, or with the body of another cell, it conducts its message as readily as a continuous filament could do—precisely as in the case of an electric wire.

This conception, founded on a most tangible anatomical basis, enables us to answer the question as to how ideas are isolated, and also, as Dr. Cajal points out, throws new light on many other mental processes. One can imagine, for example, by keeping in mind the flexible nerve prolongations, how new trains of thought may be engendered through novel associations of cells; how facility of thought or of action in certain directions is acquired through the habitual making of certain nerve–cell connections; how certain bits of knowledge may escape our memory and refuse to be found for a time because of a temporary incapacity of the nerve cells to make the proper connections, and so on indefinitely.

If one likens each nerve cell to a central telephone office, each of its filamentous prolongations to a telephone wire, one can imagine a striking analogy between the modus operandi of nervous processes and of the telephone system. The utility of new connections at the central office, the uselessness of the mechanism when the connections cannot be made, the "wires in use" that retard your message, perhaps even the crossing of wires, bringing you a jangle of sounds far different from what you desire—all these and a multiplicity of other things that will suggest themselves to every user of the telephone may be imagined as being almost ludicrously paralleled in the operations of the nervous mechanism. And that parallel, startling as it may seem, is not a mere futile imagining. It is sustained and rendered plausible by a sound substratum of knowledge of the anatomical conditions under which the central nervous mechanism exists, and in default of which, as pathology demonstrates with no less certitude, its functionings are futile to produce the normal manifestations of higher intellection.

X. THE NEW SCIENCE OF ORIENTAL ARCHAEOLOGY

HOW THE "RIDDLE OF THE SPHINX" WAS READ

Conspicuously placed in the great hall of Egyptian antiquities in the British Museum is a wonderful piece of sculpture known as the Rosetta Stone. I doubt if any other piece in the entire exhibit attracts so much attention from the casual visitor as this slab of black basalt on its telescope–like pedestal. The hall itself, despite its profusion of strangely sculptured treasures, is never crowded, but before this stone you may almost always find some one standing, gazing with more or less of discernment at the strange characters that are graven neatly across its upturned, glass–protected face. A glance at this graven surface suffices to show that three sets of inscriptions are recorded there. The upper one, occupying about one–fourth of the surface, is a pictured scroll, made up of chains of those strange outlines of serpents, hawks, lions, and so on, which are recognized, even by the least initiated, as hieroglyphics. The middle inscription, made up of lines, angles, and half–pictures, one might surmise to be a sort of abbreviated or short–hand hieroglyphic. The third or lower inscription is Greek—obviously a thing of words. If the screeds above be also made of words, only the elect have any way of proving the fact.

Fortunately, however, even the least scholarly observer is left in no doubt as to the real import of the thing he sees, for an obliging English label tells us that these three inscriptions are renderings of the same message, and that this message is a "decree of the priests of Memphis conferring divine honors on Ptolemy V. (Epiphenes), King of Egypt, B.C. 195." The label goes on to state that the upper inscription (of which, unfortunately, only part of the last dozen lines or so remains, the slab being broken) is in "the Egyptian language, in hieroglyphics, or writing of the priests"; the second inscription "in the same language is in Demotic, or the writing of the people"; and the third "the Greek language and character." Following this is a brief biography of the Rosetta Stone itself, as follows: "The stone was found by the French in 1798 among the ruins of Fort Saint Julien, near the Rosetta mouth of the Nile. It passed into the hands of the British by the treaty of Alexandria, and was deposited in the British Museum in the year 1801." There is a whole volume of history in that brief inscription—and a bitter sting thrown in, if the reader chance to be a Frenchman. Yet the facts involved could scarcely be suggested more modestly. They are recorded much more bluntly in a graven inscription on the side of the stone, which reads: "Captured in Egypt by the British Army, 1801." No Frenchman could read those words without a veritable sinking of the heart.

The value of the Rosetta Stone depended on the fact that it gave promise, even when casually inspected, of furnishing a key to the centuries–old mystery of the hieroglyphics. For two thousand years the secret of these strange markings had been forgotten. Nowhere in the world—quite as little in Egypt as elsewhere—had any man the slightest clew to their meaning; there were those who even doubted whether these droll picturings really had any specific meaning, questioning whether they were not rather vague symbols of esoteric religious import and nothing more. And it was the Rosetta Stone that gave the answer to these doubters and restored to the world a lost language and a forgotten literature.

The trustees of the museum recognized at once that the problem of the Rosetta Stone was one on which the scientists of the world might well exhaust their ingenuity, and promptly published to the world a carefully lithographed copy of the entire inscription, so that foreign scholarship had equal opportunity with the British to try at the riddle. It was an Englishman, however, who first gained a clew to the solution. This was none other than the extraordinary Dr. Thomas Young, the demonstrator of the vibratory nature of light.

Young's specific discoveries were these: (1) That many of the pictures of the hieroglyphics stand for the names of the objects actually delineated; (2) that other pictures are sometimes only symbolic; (3) that plural numbers are represented by repetition; (4) that numerals are represented by dashes; (5) that hieroglyphics may read either from the right or from the left, but always from the direction in which the animal and human figures face; (6) that proper names are surrounded by a graven oval ring, making what he called a cartouche; (7) that the cartouches of the preserved portion of the Rosetta Stone stand for the name of Ptolemy alone; (8) that the presence of a female figure after such cartouches in other inscriptions always denotes the female sex; (9) that within the cartouches the hieroglyphic symbols have a positively phonetic value, either alphabetic or syllabic; and (10) that several different characters may have the same phonetic value.

Just what these phonetic values are Young pointed out in the case of fourteen characters representing nine sounds, six of which are accepted to–day as correctly representing the letters to which he ascribed them, and the three others as being correct regarding their essential or consonant element. It is clear, therefore, that he was on the right track thus far, and on the very verge of complete discovery. But, unfortunately, he failed to take the next step, which would have been to realize that the same phonetic values which were given to the alphabetic characters within the cartouches were often ascribed to them also when used in the general text of an inscription; in other words, that the use of an alphabet was not confined to proper names. This was the great secret which Young missed and which his French successor, Jean Francois Champollion, working on the foundation that Young had laid, was enabled to ferret out.

Young's initial studies of the Rosetta Stone were made in 1814; his later publication bore date of 1819. Champollion's first announcement of results came in 1822; his second and more important one in 1824. By this time, through study of the cartouches of other inscriptions, Champollion had made out almost the complete alphabet, and the "riddle of the Sphinx" was practically solved. He proved that the Egyptians had developed a relatively complete alphabet (mostly neglecting the vowels, as early Semitic alphabets did also) centuries before the Phoenicians were heard of in history. What relation this alphabet bore to the Phoenician we shall have occasion to ask in another connection; for the moment it suffices to know that those strange pictures of the Egyptian scroll are really letters.

Even this statement, however, must be in a measure modified. These pictures are letters and something more. Some of them are purely alphabetical in character and some are symbolic in another way. Some characters represent syllables. Others stand sometimes as mere representatives of sounds, and again, in a more extended sense, as representations of things, such as all hieroglyphics doubtless were in the beginning. In a word, this is an alphabet, but not a perfected alphabet, such as modern nations are accustomed to; hence the enormous complications and difficulties it presented to the early investigators.

Champollion did not live to clear up all these mysteries. His work was taken up and extended by his pupil Rossellini, and in particular by Dr. Richard Lepsius in Germany, followed by M. Bernouf, and by Samuel Birch of the British Museum, and more recently by such well-known Egyptologists as MM. Maspero and Mariette and Chabas, in France, Dr. Brugsch, in Germany, and Dr. E. Wallis Budge, the present head of the Department of Oriental Antiquities at the British Museum. But the task of later investigators has been largely one of exhumation and translation of records rather than of finding methods.

TREASURES FROM NINEVEH

The most casual wanderer in the British Museum can hardly fail to notice two pairs of massive sculptures, in the one case winged bulls, in the other winged lions, both human–headed, which guard the entrance to the Egyptian hall, close to the Rosetta Stone. Each pair of these weird creatures once guarded an entrance to the palace of a king in the famous city of Nineveh. As one stands before them his mind is carried back over some twenty–seven intervening centuries, to the days when the "Cedar of Lebanon" was "fair in his greatness" and the scourge of Israel.

The very Sculptures before us, for example, were perhaps seen by Jonah when he made that famous voyage to Nineveh some seven or eight hundred years B.C. A little later the Babylonian and the Mede revolted against Assyrian tyranny and descended upon the fair city of Nineveh, and almost literally levelled it to the ground. But these great sculptures, among other things, escaped destruction, and at once hidden and preserved by the accumulating debris of the centuries, they stood there age after age, their very existence quite forgotten. When Xenophon marched past their site with the ill–starred expedition of the ten thousand, in the year 400 B.C., he saw only a mound which seemed to mark the site of some ancient ruin; but the Greek did not suspect that he looked upon the site of that city which only two centuries before had been the mistress of the world.

So ephemeral is fame! And yet the moral scarcely holds in the sequel; for we of to–day, in this new, undreamed–of Western world, behold these mementos of Assyrian greatness fresh from their twenty–five hundred years of entombment, and with them records which restore to us the history of that long–forgotten people in such detail as it was not known to any previous generation since the fall of Nineveh. For two thousand five hundred years no one saw these treasures or knew that they existed. One hundred generations of men came and went without once pronouncing the name of kings Shalmaneser or Asumazirpal or Asurbanipal. And to–day, after these centuries of oblivion, these names are restored to history, and, thanks to the character of their monuments, are assured a permanency of fame that can almost defy time itself. It would be nothing strange, but rather in keeping with their previous mutations of fortune, if the names of Asurnazirpal and Asurbanipal should be familiar as household words to future generations that have forgotten the existence of an Alexander, a Caesar, and a Napoleon. For when Macaulay's prospective New Zealander explores the ruins of the British Museum the records of the ancient Assyrians will presumably still be there unscathed, to tell their story as they have told it to our generation, though every manuscript and printed book may have gone the way of fragile textures.

But the past of the Assyrian sculptures is quite necromantic enough without conjuring for them a necromantic future. The story of their restoration is like a brilliant romance of history. Prior to the middle of this century the inquiring student could learn in an hour or so all that was known in fact and in fable of the renowned city of Nineveh. He had but to read a few chapters of the Bible and a few pages of Diodorus to exhaust the important literature on the subject. If he turned also to the pages of Herodotus and Xenophon, of Justin and Aelian, these served chiefly to confirm the suspicion that the Greeks themselves knew almost nothing more of the history of their famed Oriental forerunners. The current fables told of a first King Ninus and his wonderful queen Semiramis; of Sennacherib the conqueror; of the effeminate Sardanapalus, who neglected the warlike ways of his ancestors but perished gloriously at the last, with Nineveh itself, in a self−imposed holocaust. And that was all. How much of this was history, how much myth, no man could say; and for all any one suspected to the contrary, no man could ever know. And to−day the contemporary records of the city are before us in such profusion as no other nation of antiquity, save Egypt alone, can at all rival. Whole libraries of Assyrian books are at hand that were written in the seventh century before our era. These, be it understood, are the original books themselves, not copies. The author of that remote time appeals to us directly, hand to eye, without intermediary transcriber. And there is not a line of any Hebrew or Greek manuscript of a like age that has been preserved to us; there is little enough that can match these ancient books by a thousand years. When one reads Moses or Isaiah, Homer, Hesiod, or Herodotus, he is but following the transcription—often unquestionably faulty and probably never in all parts perfect—of successive copyists of later generations. The oldest known copy of the Bible, for example, dates probably from the fourth century A.D., a thousand years or more after the last Assyrian records were made and read and buried and forgotten.

There was at least one king of Assyria—namely, Asurbanipal, whose palace boasted a library of some ten thousand volumes—a library, if you please, in which the books were numbered and shelved systematically, and classified and cared for by an official librarian. If you would see some of the documents of this marvellous library you have but to step past the winged lions of Asurnazirpal and enter the Assyrian hall just around the corner from the Rosetta Stone. Indeed, the great slabs of stone from which the lions themselves are carved are in a sense books, inasmuch as there are written records inscribed on their surface. A glance reveals the strange characters in which these records are written, graven neatly in straight lines across the stone, and looking to casual inspection like nothing so much as random flights of arrow–heads. The resemblance is so striking that this is sometimes called the arrow–head character, though it is more generally known as the wedge or cuneiform character. The inscriptions on the flanks of the lions are, however, only makeshift books. But the veritable books are no farther away than the next room beyond the hall of Asurnazirpal. They occupy part of a series of cases placed down the centre of this room. Perhaps it is not too much to speak of this collection as the most extraordinary set of documents of all the rare treasures of the British Museum, for it includes not books alone, but public and private letters, business announcements, marriage contracts—in a word, all the species of written records that enter into the every–day life of an intelligent and cultured community.

But by what miracle have such documents been preserved through all these centuries? A glance makes the secret evident. It is simply a case of time–defying materials. Each one of these Assyrian documents appears to be, and in reality is, nothing more or less than an inscribed fragment of brick, having much the color and texture of a weathered terra–cotta tile of modern manufacture. These slabs are usually oval or oblong in shape, and from two or three to six or eight inches in length and an inch or so in thickness. Each of them was originally a portion of brick–clay, on which the scribe indented the flights of arrowheads with some sharp–cornered instrument, after which the document was made permanent by baking. They are somewhat fragile, of course, as all bricks are, and many of them have been more or less crumbled in the destruction of the palace at Nineveh; but to the ravages of mere time they are as nearly invulnerable as almost anything in nature. Hence it is that these records of a remote civilization have been preserved to us, while the similar records of such later civilizations as the Grecian have utterly perished, much as the flint implements of the cave–dweller come to us unchanged, while the iron implements of a far more recent age have crumbled away.

HOW THE RECORDS WERE READ

After all, then, granted the choice of materials, there is nothing so very extraordinary in the mere fact of preservation of these ancient records. To be sure, it is vastly to the credit of nineteenth–century enterprise to have searched them out and brought them back to light. But the real marvel in connection with them is the fact that nineteenth–century scholarship should have given us, not the material documents themselves, but a knowledge of their actual contents. The flight of arrow–heads on wall or slab or tiny brick have surely a meaning; but how shall we guess that meaning? These must be words; but what words? The hieroglyphics of the Egyptians were mysterious enough in all conscience; yet, after all, their symbols have a certain suggestiveness, whereas there is nothing that seems to promise a mental leverage in the unbroken succession of these cuneiform dashes. Yet the Assyrian scholar of to–day can interpret these strange records almost as readily and as surely as the classical scholar interprets a Greek manuscript. And this evidences one of the greatest triumphs of nineteenth–century scholarship, for within almost two thousand years no man has lived, prior to our century, to whom these strange inscriptions would not have been as meaningless as they are to the most casual stroller who looks on them with vague wonderment here in the museum to–day. For the Assyrian language, like the Egyptian, was veritably a dead language; not, like Greek and Latin, merely passed from practical every–day use to the closet of the scholar, but utterly and absolutely forgotten by all the world. Such being the case, it is nothing less than marvellous that it should have been restored.

It is but fair to add that this restoration probably never would have been effected, with Assyrian or with Egyptian, had the language in dying left no cognate successor; for the powers of modern linguistry, though great, are not actually miraculous. But, fortunately, a language once developed is not blotted out in toto; it merely outlives its usefulness and is gradually supplanted, its successor retaining many traces of its origin. So, just as Latin, for example, has its living representatives in Italian and the other Romance tongues, the language of Assyria is represented by cognate Semitic languages. As it chances, however, these have been of aid rather in the later stages of Assyrian study than at the very outset; and the first clew to the message of the cuneiform writing came through a slightly different channel.

Curiously enough, it was a trilingual inscription that gave the clew, as in the case of the Rosetta Stone, though with very striking difference withal. The trilingual inscription now in question, instead of being a small, portable monument, covers the surface of a massive bluff at Behistun in western Persia. Moreover, all three of its inscriptions are in cuneiform characters, and all three are in languages that at the beginning of our century were absolutely unknown. This inscription itself, as a striking monument of unknown import, had been seen by successive generations. Tradition ascribed it, as we learn from Ctesias, through Diodorus, to the fabled Assyrian queen Semiramis. Tradition was quite at fault in this; but it is only recently that knowledge has availed to set it right. The inscription, as is now known, was really written about the year 515 B.C., at the instance of Darius I., King of Persia, some of whose deeds it recounts in the three chief languages of his widely scattered subjects.

The man who at actual risk of life and limb copied this wonderful inscription, and through interpreting it became the veritable "father of Assyriology," was the English general Sir Henry Rawlinson. His feat was another British triumph over the same rivals who had competed for the Rosetta Stone; for some French explorers had been sent by their government, some years earlier, expressly to copy this strange record, and had reported that it was impossible to reach the inscription. But British courage did not find it so, and in 1835 Rawlinson scaled the dangerous height and made a paper cast of about half the inscription. Diplomatic duties called him away from the task for some years, but in 1848 he returned to it and completed the copy of all parts of the inscription that have escaped the ravages of time. And now the material was in hand for a new science, which General Rawlinson himself soon, assisted by a host of others, proceeded to elaborate.

The key to the value of this unique inscription lies in the fact that its third language is ancient Persian. It appears that the ancient Persians had adopted the cuneiform character from their western neighbors, the Assyrians, but in so doing had made one of those essential modifications and improvements which are scarcely possible to accomplish except in the transition from one race to another. Instead of building with the arrow–head a multitude of syllabic characters, including many homophones, as had been and continued to be the custom with the Assyrians, the Persians selected a few of these characters and ascribed to them phonetic values that were almost purely alphabetic. In a word, while retaining the wedge as the basal stroke of their script, they developed an alphabet, making the last wonderful analysis of phonetic sounds which even to this day has escaped the Chinese, which the Egyptians had only partially effected, and which the Phoenicians were accredited by the Greeks with having introduced to the Western world. In addition to this all–essential step, the Persians had introduced the minor but highly convenient custom of separating the words of a sentence from one another by a particular mark, differing in this regard not only from the Assyrians and Egyptians, but from the early Greek scribes as well.

Thanks to these simplifications, the old Persian language had been practically restored about the beginning of the nineteenth century, through the efforts of the German Grotefend, and further advances in it were made just at this time by Renouf, in France, and by Lassen, in Germany, as well as by Rawlinson himself, who largely solved the problem of the Persian alphabet independently. So the Persian portion of the Behistun inscription could be at least partially deciphered. This in itself, however, would have been no very great aid towards the restoration of the languages of the other portions had it not chanced, fortunately, that the inscription is sprinkled with proper names. Now proper names, generally speaking, are not translated from one language to another, but transliterated as nearly as the genius of the language will permit. It was the fact that the Greek word Ptolemaics was transliterated on the Rosetta Stone that gave the first clew to the sounds of the Egyptian characters. Had the upper part of the Rosetta Stone been preserved, on which, originally, there were several other names, Young would not have halted where he did in his decipherment.

But fortune, which had been at once so kind and so tantalizing in the case of the Rosetta Stone, had dealt more gently with the Behistun inscriptions; for no fewer than ninety proper names were preserved in the Persian portion and duplicated, in another character, in the Assyrian inscription. A study of these gave a clew to the sounds of the Assyrian characters. The decipherment of this character, however, even with this aid, proved enormously difficult, for it was soon evident that here it was no longer a question of a nearly perfect alphabet of a few characters, but of a syllabary of several hundred characters, including many homophones, or different forms for representing the same sound. But with the Persian translation for a guide on the one hand, and the Semitic languages, to which family the Assyrian belonged, on the other, the appalling task was gradually accomplished, the leading investigators being General Rawlinson, Professor Hincks, and Mr. Fox–Talbot, in England, Professor Jules Oppert, in Paris, and Professor Julian Schrader, in Germany, though a host of other scholars soon entered the field.

This great linguistic feat was accomplished about the middle of the nineteenth century. But so great a feat was it that many scholars of the highest standing, including Joseph Erneste Renan, in France, and Sir G. Cornewall Lewis, in England, declined at first to accept the results, contending that the Assyriologists had merely deceived themselves by creating an arbitrary language. The matter was put to a test in 1855 at the suggestion of Mr. Fox–Talbot, when four scholars, one being Mr. Talbot himself and the others General Rawlinson, Professor Hincks, and Professor Oppert, laid before the Royal Asiatic Society their independent interpretations of a hitherto untranslated Assyrian text. A committee of the society, including England's greatest historian of the century, George Grote, broke the seals of the four translations, and reported that they found them unequivocally in accord as regards their main purport, and even surprisingly uniform as regards the phraseology of certain passages—in short, as closely similar as translations from the obscure texts of any difficult language ever are. This decision gave the work of the Assyriologists official status, and the reliability of their method has never since been in question. Henceforth Assyriology was an established science.

APPENDIX

REFERENCE–LIST

CHAPTER I. MODERN DEVELOPMENT OF THE PHYSICAL SCIENCES

[1] Robert Boyle, Philosophical Works (3 vols.). London, 1738.

CHAPTER II. THE BEGINNINGS OF MODERN CHEMISTRY

[1] For a complete account of the controversy called the "Water Controversy," see The Life of the Hon. Henry Cavendish, by George Wilson, M.D., F.R.S.E. London, 1850.

[2] Henry Cavendish, in Phil. Trans. for 1784, P. 119.

[3] Lives of the Philosophers of the Time of George III., by Henry, Lord Brougham, F.R.S., p. 106. London, 1855.

[4] Experiments and Observations on Different Kinds of Air, by Joseph Priestley (3 vols.). Birmingham, 790, vol. II, pp. 103–107.

[5] Lectures on Experimental Philosophy, by Joseph Priestley, lecture IV., pp. 18, ig. J. Johnson, London, 1794.

[6] Translated from Scheele's Om Brunsten, eller Magnesia, och dess Egenakaper. Stockholm, 1774, and published as Alembic Club Reprints, No. 13, 1897, p. 6.

[7] According to some writers this was discovered by Berzelius.

[8] Histoire de la Chimie, par Ferdinand Hoefer. Paris, 1869, Vol. CL, p. 289.

[9] Elements of Chemistry, by Anton Laurent Lavoisier, translated by Robert Kerr, p. 8. London and Edinburgh, 1790.

[10] Ibid., pp. 414–416.

CHAPTER III. CHEMISTRY SINCE THE TIME OF DALTON

[1] Sir Humphry Davy, in Phil. Trans., Vol. VIII.

CHAPTER IV. ANATOMY AND PHYSIOLOGY IN THE EIGHTEENTH CENTURY

[1] Baas, History of Medicine, p. 692.

[2] Based on Thomas H. Huxley's Presidential Address to the British Association for the Advancement of Science, 1870.

[3] Essays on Digestion, by James Carson. London, 1834, p. 6.

[4] Ibid., p. 7.

[5] John Hunter, On the Digestion of the Stomach after Death, first edition, pp. 183–188.

[6] Erasmus Darwin, The Botanic Garden, pp. 448–453. London, 1799.

CHAPTER V. ANATOMY AND PHYSIOLOGY IN THE NINETEENTH CENTURY

[1] Baron de Cuvier's Theory of the Earth. New York, 1818, p. 123.

[2] On the Organs and Mode of Fecundation of Orchidex and Asclepiadea, by Robert Brown, Esq., in Miscellaneous Botanical Works. London, 1866, Vol. I., pp. 511–514.

[3] Justin Liebig, Animal Chemistry. London, 1843, p. 17f.

CHAPTER VI. THEORIES OF ORGANIC EVOLUTION

[1] "Essay on the Metamorphoses of Plants," by Goethe, translated for the present work from Grundriss einer Geschichte der Naturwissenschaften, by Friederich Dannemann (2 vols.). Leipzig, 1896, Vol. I., p. 194.

[2] The Temple of Nature, or The Origin of Society, by Erasmus Darwin, edition published in 1807, p. 35.

[3] Baron de Cuvier, Theory of the Earth. New York, 1818, p.74. (This was the introduction to Cuvier's great work.)

[4] Robert Chambers, Explanations: a sequel to Vestiges of Creation. London, Churchill, 1845, pp. 148–153.

CHAPTER VII. EIGHTEENTH–CENTURY MEDICINE

[1] Condensed from Dr. Boerhaave's Academical Lectures on the Theory of Physic. London, 1751, pp. 77, 78. Boerhaave's lectures were published as Aphorismi de cognoscendis et curandis Morbis, Leyden, 1709. On this book Van Swieten wrote commentaries filling five volumes. Another very celebrated work of Boerhaave is his Institutiones et Experimenta Chemic, Paris, 1724, the germs of this being given as a lecture on his appointment to the chair of chemistry in the University of Leyden in 1718.

[2] An Inquiry into the Causes and Effects of the Variola Vaccine, etc., by Edward Jenner, M.D., F.R.S., etc. London, 1799, pp. 2–7. He wrote several other papers, most of which were communications to the Royal Society. His last publication was, On the Influence of Artificial Eruptions in Certain Diseases (London, 1822), a subject to which he had given much time and study.

CHAPTER VIII. NINETEENTH–CENTURY MEDICINE

[1] In the introduction to Corvisart's translation of Avenbrugger's work. Paris, 1808.

[2] Laennec, Traite d'Auscultation Mediate. Paris, 1819. This was Laennec's chief work, and was soon translated into several different languages. Before publishing this he had written also, Propositions sur la doctrine midicale d'Hippocrate, Paris, 1804, and Memoires sur les vers visiculaires, in the same year.

[3] Researches, Chemical and Philosophical, chiefly concerning Nitrous Oxide or Dephlogisticated Nitrous Air and its Respiration, by Humphry Davy. London, 1800, pp. 479–556.

[4] Ibid.

[5] For accounts of the discovery of anaesthesia, see Report of the Board of Trustees of the Massachusetts General Hospital, Boston, 1888. Also, The Ether Controversy: Vindication of the Hospital Reports of 1848, by N. L Bowditch, Boston, 1848. An excellent account is given in Littell's Living Age, for March, 1848, written by R. H. Dana, Jr. There are also two Congressional Reports on the question of the discovery of etherization, one for 1848, the other for 11852.

[6] Simpson made public this discovery of the anaesthetic properties of chloroform in a paper read before the Medico–Chirurgical Society of Edinburgh, in March, 1847, about three months after he had first seen a surgical operation performed upon a patient to whom ether had been administered.

[7] Louis Pasteur, Studies on Fermentation. London, 1870.

[8] Louis Pasteur, in Comptes Rendus des Sciences de L'Academie des Sciences, vol. XCII., 1881, pp. 429–435.

CHAPTER IX. THE NEW SCIENCE OF EXPERIMENTAL PSYCHOLOGY

[1] Bell's communications were made to the Royal Society, but his studies and his discoveries in the field of anatomy of the nervous system were collected and published, in 1824, as An Exposition of the Natural System of Nerves of the Human Body: being a Republication of the Papers delivered to the Royal Society on the Subject of the Nerves.

[2] Marshall Hall, M.D., F.R.S.L., On the Reflex Functions of the Medulla Oblongata and the Medulla Spinalis, in Phil. Trans. of Royal Soc., vol. XXXIII., 1833.

Printed in the United States
40784LVS00007B/79